The Object Lessons series achieves something very close to magic: the books take ordinary—even banal—objects and animate them with a rich history of invention, political struggle, science, and popular mythology. Filled with fascinating details and conveyed in sharp, accessible prose, the books make the everyday world come to life. Be warned: once you've read a few of these, you'll start walking around your house, picking up random objects, and musing aloud: "I wonder what the story is behind this thing?"'

Steven Johnson, author of *Where Good Ideas Come From* and *How We Got to Now*

Object Lessons describes themselves as 'short, beautiful books,' and to that, I'll say, amen. . . . If you read enough Object Lessons books, you'll fill your head with plenty of trivia to amaze and annoy your friends and loved ones—caution recommended on pontificating on the objects surrounding you. More importantly, though … they inspire us to take a second look at parts of the everyday that we've taken for granted. These are not so much lessons about the objects themselves, but opportunities for self-reflection and storytelling. They remind us that we are surrounded by a wondrous world, as long as we care to look.'

John Warner, *The Chicago Tribune*

'Besides being beautiful little hand-sized objects themselves, showcasing exceptional writing, the wonder of these books is that they exist at all . . . Uniformly excellent, engaging, thought-provoking, and informative.'

Jennifer Bort Yacovissi, *Washington Independent Review of Books*

'. . . edifying and entertaining . . . perfect for slipping in a pocket and pulling out when life is on hold.'

Sarah Murdoch, *Toronto Star*

'For my money, Object Lessons is the most consistently interesting nonfiction book series in America.'

Megan Volpert, *PopMatters*

'Though short, at roughly 25,000 words apiece, these books are anything but slight.'

Marina Benjamin, *New Statesman*

'[W]itty, thought-provoking, and poetic . . . These little books are a page-flipper's dream.'

John Timpane, *The Philadelphia Inquirer*

The joy of the series, of reading *Remote Control, Golf Ball, Driver's License, Drone, Silence, Glass, Refrigerator, Hotel,* and *Waste* (more titles are listed as forthcoming) in quick succession, lies in encountering the various turns through which each of their authors has been put by his or her object. As for Benjamin, so for the authors of the series, the object predominates, sits squarely center stage, directs the action. The object decides the genre, the chronology, and the limits of the study. Accordingly, the author has to take her cue from the *thing* she chose or that chose her. The result is a wonderfully uneven series of books, each one a *thing* unto itself.'

Julian Yates, *Los Angeles Review of Books*

The Object Lessons series has a beautifully simple premise. Each book or essay centers on a specific object. This can be mundane or unexpected, humorous or politically timely. Whatever the subject, these descriptions reveal the rich worlds hidden under the surface of things.'

Christine Ro, *Book Riot*

. . . a sensibility somewhere between Roland Barthes and Wes Anderson.'

Simon Reynolds, author of *Retromania: Pop Culture's Addiction to Its Own Past*

OBJECTLESSONS

A book series about the hidden lives of ordinary things.

Series Editors:

Ian Bogost and Christopher Schaberg

In association with

BOOKS IN THE SERIES

Alarm

ALICE BENNETT

BLOOMSBURY ACADEMIC
NEW YORK • LONDON • OXFORD • NEW DELHI • SYDNEY

BLOOMSBURY ACADEMIC
Bloomsbury Publishing Inc
1385 Broadway, New York, NY 10018, USA
50 Bedford Square, London, WC1B 3DP, UK
29 Earlsfort Terrace, Dublin 2, Ireland

BLOOMSBURY, BLOOMSBURY ACADEMIC and the Diana logo are
trademarks of Bloomsbury Publishing Plc

First published in the United States of America 2023

Library of Congress Cataloging-in-Publication Data

Names: Bennett, Alice, 1982- author.
Title: Alarm/Alice Bennett.
Description: New York, NY: Bloomsbury Academic, 2023. | Series: Object lessons | Includes
bibliographical references and index. | Summary: "A book that invites you to listen more closely to
the beepers, bells, and buzzers that have shaped modern minds"– Provided by publisher.
Identifiers: LCCN 2022022667 (print) | LCCN 2022022668 (ebook) | ISBN 9781501375576
(paperback) | ISBN 9781501375583 (epub) | ISBN 9781501375590 (pdf) | ISBN 9781501375606
Subjects: LCSH: Noise–Psychological aspects. | Fire alarms. | False alarms. |
Emergency communication systems.
Classification: LCC BF205.N6 B45 2023 (print) | LCC BF205.N6 (ebook) |
DDC 152.1/5–dc23/eng/20220825
LC record available at https://lccn.loc.gov/2022022667
LC ebook record available at https://lccn.loc.gov/2022022668

ISBN: PB: 978-15013-7557-6
ePDF: 978-1-5013-7559-0
eBook: 978-1-5013-7558-3

Series: Object Lessons

Typeset by Deanta Global Publishing Services, Chennai, India
Printed and bound in the United States of America

To find out more about our authors and books visit www.bloomsbury.com and
sign up for our newsletters.

CONTENTS

FIGURES

INTRODUCTION

More than just bells and whistles, alarms are objects that have shaped some of the most fundamental parts of modern life. Sleeping and waking, security and danger, concentration and distraction: all are invigilated by the alarm. And from the watchtower to the smartwatch, the history of alarms is also a history of safety, surveillance, work, automation, emotion, and attention. The blaring police siren and the beeping smoke alarm, the chiming cathedral bell and the buzzing phone alert carry messages about what we value and how we respond to the world around us.

This book is about the attention-grabbing but unobtrusive alarm as an object that organises and administers feelings. Alarms direct rest and arousal, calm and alarm, in order to manage our alertness and allow us to unburden ourselves of the responsibility of ever-watchful, vigilant attention. We depend on them for our safety, our sleep, and sometimes even for our self-control. But there is a contradiction at work in the alarm, which guarantees relief from the overwhelming attentive work of constant vigilance only by seizing attention with great urgency. When activated, alarms interrupt, disrupt and disturb. Alarms *alarm*. They wake us up and force us to turn our attention to things we've preferred to ignore.

In its function as an alertness-management device, the alarm offers a respite from awareness. Using a kitchen timer means that it's possible to boil an egg and do the washing up at the same time. Installing a smoke alarm enables sounder sleep, with fewer worries about housefires. Ultimately, the alarm remains alert while we immerse ourselves in absorbed concentration or oblivious distraction. Alarms should therefore be understood as objects that augment attention but also enable certain comfortable and necessary states of unconsciousness, ignorance, or inattentiveness.

One way to conceptualise the alarm as a manager of attention is to understand it as an example of the prosthesis, those 'auxiliary organs' with which Freud describes people equipping themselves as part of a fantasy of perfectibility.[1] Understood in this way, the alarm is a prosthetic of vigilance; an artificial supplement to and substitute for our own alertness or attention. Alarms represent a specific kind of watchful, unfailing attention – a perfect and godlike omniscience – which, by implication, we lack ourselves. The alarm supplements fallible human attention with a technological fantasy of always-on, inexhaustible vigilance. Nevertheless, as Freud observes, these technological additions cause 'a good deal of trouble on occasion'.[2] Inevitably, we are unhappy with how we've augmented ourselves.

The alarm's malfunctions or dysfunctions – false alarms, alarm fatigue, or any of the mechanisms of bypassing and over-riding and disabling alarms – are the places we see most clearly the potential for trouble that Freud identifies in the prosthesis. As anyone who's ever hated their alarm clock or their neighbour's car alarm will recognise, it's not uncommon to feel resentment or irritation towards alarms. The alarm promises a rationalised and optimised allocation of states of alertness, but the interface between alarm attention and our own attention is not always an easy one. The places where the technology irritates us, or extends our alertness in ways that are too dull or too acute: these are the places that are often the most revealing of our relationship with the technology we've created.

The hopes placed in the alarm as an extension of attention – the guarantee that nothing should escape our notice, that monitoring ensures not just safety but also efficiency, that attention is perfectible – therefore suggest that this object represents profound anxiety about the limits of human attention. Nevertheless, even this conception of attention as deficient is a product of the norms of neurotypicality. If we start from the position that attention is enormously diverse in its forms, and may overspill any attempts to measure its span or capacity, then it becomes more difficult to think of alarms as only correcting a shortfall of attention. For many people (especially those who are autistic or who have ADHD) alarms are an assistive technology that acknowledges and enables differences in attention rather than amending a deficiency. Alarms allow for the flourishing of states of hyperfocus and absorption, and the pleasurable experience of what positive psychology calls 'flow'.[3] The alarm is therefore two things at once: an object that can shape attention in ways that make the interface between us and the world easier, safer and happier, and an object that administers our attention in ways that can feel uncomfortable, demanding, and burdensome. As this book unfolds, we will see both aspects of the alarm represented, sometimes at the same time.

* * *

The word 'alarm' holds inside it a command. Its origin is the imperative, 'To arms!', from the Italian *all'arme*.[4] A functional and expansive definition of the term 'alarm' would therefore have two parts: a call to attention and a demand for an urgent response. It's the second part of this definition which distinguishes the alarm from other calls to attention such as alerts or notifications. An alert, like an alarm, draws attention to itself, but may or may not require action. The word 'alert' also has spatial connotations trailing in its etymology that are absent from the word 'alarm'. 'Alert' also comes from Italian, originating with the phrase *all'erta*, meaning 'on the watch', or 'on the lookout'. To be *all'erta* is to physically occupy the 'high point' or promontory, and therefore to have an elevated viewpoint.[5] At the same time, the implication is that to be properly *all'erta* requires a particular state of perception and arousal; a watchman may be up on the promontory above his Italian town, but that's no use if he's dozing or daydreaming. To be alert, therefore, is to be in a state of readiness, primed to spring into action and call *all'arme* to others. The two terms' etymologies bind them together and suggest gradations and distinctions within these calls to attention, as well as forms of shared or communal attention that interact with each other.

Theorists of technology identify alarms, alerts, warnings and notifications as having the property of being 'designed-to-occur'.[6] This means that the crucial, defining feature of an alarm is that it *happens* – it's an event that occurs in time. All alarms are responsive events which interrupt and disrupt by drawing attention to themselves. In the context of automated systems, the alarm is the indication that human intervention is required. My washing machine, for instance, can run a load of laundry, regulating the temperature, water input and output and spin-speed automatically, and adding in soap and fabric conditioner at pre-programmed times in the cycle. But it still beeps an alarm when the laundry is done so that I don't leave the wet things mouldering in the drum. On a much larger scale, an airliner or a nuclear power plant can be automated to run itself, barring those alarmed processes which require operator intervention. In an age of automation, then, the alarm's call has proliferated. The counterpart to the automated system that watches itself is the alarm that calls for a user's attention only when absolutely necessary. The term 'alert' has taken on a new meaning in an age of automated systems, too, and has come to represent those forms of informational message or notification that, by definition, don't require an immediate response. If an alarm is a raucous klaxon or a disruptive ringing, an alert is the soft ping of a calendar reminder or the haptic nudge of a smartwatch. In the twenty-first century a whole life can be gradually prodded into shape by ever more ambient structures of notification.

Within this framework of the 'designed-to-occur', the alarm is an event designed at one time to occur at another. Moreover, the alarm only has any meaning at all if the people who are called up by it have a previous understanding that the designed-to-occur event has been designed to occur. Drills, rehearsals, alarm testing, evacuation procedures, disaster simulations: alarms accrue processes that make sure the attention they invoke can be converted into action. The alarm should be a surprise (or what was the point of it taking over our own vigilance?) but it should not be completely unanticipated. In many cases, the alarm represents the siphoning off of attention from one time to another, as it takes over attention during the vigilance phase and then demands it urgently when it's activated. Alarms, therefore, are future-oriented. They let us drop our guard, rest, climb down from anxious vigilance, turn our attention elsewhere, but with the caveat that we will be called back to high alert at some point in the future.

*　*　*

It's clear that alarms are not just technological objects, but social and cultural objects too. They give rise to and depend upon practices that construct conventions for their use and acknowledgement of their limitations. This book is not intended to give anything like a comprehensive history of the science or technology of alarms. Instead, it analyses alarms in art, media and culture to try to understand what meanings they carry. It proceeds by interpreting examples of cultural objects – books, plays, films, songs, fables, pictures, adverts, instruction manuals – which represent and reimagine aspects of the alarm. We will encounter alarms in 90s hip hop and Hollywood heist movies, in dystopian novels and in the art of the modernist avant-garde. Each of these examples reveals aspects of the alarm's social and cultural context. Along with sending out its call to arms, to action, and to attention, the alarm sends out other signals that deserve careful interpretation.

Let's look at a very early example: *Agamemnon*, the first play of Aeschylus's *Oresteia*. In this tragedy from the 5th century BCE, we see both a fascination with the mechanics of alarm systems and an exploration, primarily through the character of Cassandra, of the social processes that surround the raising of alarm. The play stages the story of Cassandra who, cursed with the gift of prophecy, predicts the tragedy of her own murder and the murder of King Agamemnon at the hands of his wife, Clytemnestra. Cassandra raises an alarm and her alarm is dismissed as false.

The play begins with a watchman, who tells us that he has sat on the roof of Agamemnon's palace for a year of sleepless nights looking out for the lighting of the *phryctoriae* or signal-fires that will announce the fall of Troy and the triumphant return of the king. Soon, he sees the alarm of the signal-fire off in the distance and rushes off to tell the queen. However, the old men of the city refuse to believe Queen Clytemnestra when she tells them the news, thinking that she has heard it from a soothsayer or in a dream. To convince them, she gives an extraordinary, vivid speech describing how her alarm system works, listing every place from Troy to Argos where runners have passed a torch in relay from beacon to beacon.[7] Over 30 lines she evokes the keen vigilance of every watcher at every beacon – all, she stresses, at her directive. This is a boast about her planning, but also a reminder of her power, which she bears in relay, like a torch passed on from her husband. Signal beacons like the one described in the play would more commonly have been used as an early-warning system for a call to arms in case of attack. However, as the audience might realise with the benefit of dramatic irony, this is not an alarm repurposed for peace; the beacon really is a signal for Clytemnestra to arm herself, to get ready to execute Agamemnon and avenge his ritual killing of their daughter.

Aeschylus's version of one of the oldest stories in European literature therefore begins with a compelling and dramatic description of a beautifully functioning alarm system. The whole city of Argos seems to have lost faith in prophecies after the last advice they were given saw all their sons sent off to Troy, but Clytemnestra's technological solution essentially bypasses supernatural advisors and allows her to gather information through an early-warning system that seems like a prophecy. Clytemnestra's alarm beacons are therefore established in parallel to and as a disavowal of prophecies and portents. Like the beacons, Cassandra too conveys alarms from Troy when she is brought to Argos as a war trophy, enslaved by Agamemnon. Her cries, when she is overpowered by supernatural foresight, echo the shouts from the watchman at the beginning of the play, inviting a direct comparison between these two acts of alarm-raising.

What we see in *Agamemnon* is both the representation of a particular alarm technology and of the dynamics of gender and social position involved in raising the alarm. The play's form allows for these two features – the technological and the social – to come into dialogue with each other, as Aeschylus invites comparisons between the watchman and Cassandra, and between the signal-fires and divinely inspired prophecy.

In Aeschylus's play, it's clear that audiences are invited to feel for and with Cassandra as she walks through the palace doors to her fate, and therefore to recognise that being audience to a tragedy places us in alliance with her. Because, like us, Cassandra knows what is going to happen in this famous story she almost begins to seem like an audience member who has jumped on stage to raise the alarm and intervene in the plot. At the same time, watching the play and knowing the fate of the characters while being powerless to change the outcome of events, we become an audience of Cassandras. The play uses an audience's foreknowledge of events to place them, dramatically, into the position of a character who could raise an alarm but cannot stop the events onstage unfolding.

The figure of the watchman who opens the play also offers a reflexive invitation for the audience to examine their own action of alert watching. The watchman is a figure who recurs in dramatic forms, from the opening of Shakespeare's *Hamlet* to Vladimir and Estragon, watching and waiting for Godot in Beckett's play. He is a figure perhaps closely related to the shepherd in pastoral poetry – someone whose job consists of looking out for threats and raising the alarm. Audience-members too are a variety of watchman; they are on the alert, looking out (sometimes even from *all'erte* or above) for something that, like an alarm, is 'designed to occur'. Reading *Agamemnon* with alarms in mind therefore reveals how the play uses its own dramatic form to organise self-reflexive vigilance, alertness, foresight, and other modes of monitoring attention within an audience, and encourages us to take up positions within the social and emotional dynamics of the raising of alarm.

* * *

While Aeschylus's play depicts a specific alarm technology in its extraordinary staging of the signal beacon early-warning system, it also offers insight into more general principles of alarm-raising. In what follows in this book, we will encounter alarms in different time periods and serving many different purposes. From ancient water clocks to nineteenth-century burglar alarms, from mid-century smoke detectors to contemporary activity monitors, the history of alarms is rich and revealing of the societies from which these objects come, but in each case the cultural representations also accentuate some common features across times and places.

Underlying my argument in this book is the premise that all alarms have something in common: they both alleviate and enable certain states of attention through their monitoring vigilance. They can administer attention and also liberate it. They maintain structures of preparedness or anticipation that cast them forward into the future. And, as a call to arms, they require action as well as attention. But if I've been asserting that the alarm mediates between opposing affects or states (calm/alarm, sleep/wake, inertness/alertness), surely I have to acknowledge a difference between the feelings invoked by a microwave ping and a disaster siren? There is undoubtedly a difference in scale, but in general alarms are more alike than they seem. They are framed by cultural messages that offer a promise of protection from uncomfortable states of arousal – because alarms give us permission to ignore things. Thus they also entrench a preference for staying on the side of calm, inertness or even ignorance, rather than waking up to the alarm's urgent demands.

In representing alarms, writers, visual artists, film-makers and musicians not only describe those aspects of the alarm's logic. As we see in the chapters that follow, artists also take up and make use of the alarm as an object that allows them to explore specific political, artistic, and historical questions.

And, as in the example of Aeschylus's *Agamemnon,* the alarm's call to attention and to action also finds a response in aesthetic strategies. In an argument which will become more explicit towards the end of this book, I will make the case that art doesn't just reveal aspects of alarms and their processes, but also reimagines them. Ultimately, art that responds to the alarm sounds out and re-sounds the alarm's central contradictions, and prepares us to answer the alarm's call to attention and to action in new ways.

1 CLOCK

In January 1919 the painter Francis Picabia was staying at a hotel in Switzerland after returning to Europe from war work in the United States.[1] He was visited there by Tristan Tzara and Hans Arp, two members of Zurich Dada who had seen Picabia's newly published volume of poems and drawings celebrating the art of the machine and were keen to meet him. When they arrived at Picabia's hotel room they found him in the process of dismantling an alarm clock. Arp gives an account of this memorable first impression, describing the artist disassembling the alarm clock to expose its 'secret parts' (*geheimen Teilchen*) and whipping out the *Uhrfeder*, or mainspring, with a gesture of triumph.[2] One by one, Picabia began to press the pieces onto an ink pad and apply the inked-up mechanism of the clock onto a sheet of paper. Picabia was printing his alarm clock.[3]

The result is something which looks like an instructional diagram or user's manual. It's an impression that's heightened by the presence of the title – *Réveil Matin*, 'alarm clock' – as a descriptive label written on the image itself (fig. 1). We look

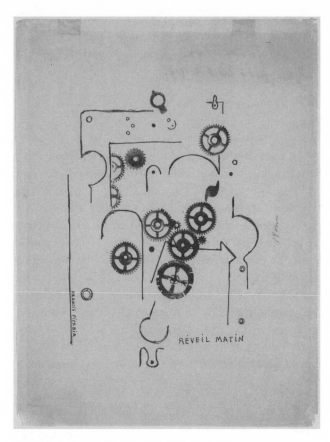

FIGURE 1 Francis Picabia, *Réveil Matin*, 1919.

down at the workings of the clock from above and laid out flat, like a clockmaker contemplating a repair.

The printed clock therefore works as an arresting image of stopped time, a paused moment preserved now for over a hundred years. Nevertheless, as we try to connect the pieces of the mechanism in the print, attempting to trace how the cogs bite or how the workings might run, it becomes clear that something is wrong: parts are missing, the lines that Picabia has drawn between the pieces don't quite connect, and it's impossible to follow how the clockwork would actually function. The hand-drawn additions bridge some of the gaps and keep the pieces together, but the longer you spend looking at *Réveil Matin*, the more obvious it becomes that this is not a blueprint or an exploded view. The clock can't possibly work. One version of an impression – in which a print exactly preserves the printed object – has been superseded by another: the artist's 'impression' of the alarm clock.

Picabia would go on to produce another diagrammatic image of an alarm clock, this time an alarm clock that was also a lesson in art history. In this piece, titled *Mouvement Dada* (1919) – a fairly corny pun that associates a horological movement with an avant-garde movement – we see a diagram of the mechanism of an alarm clock, complete with battery, transformer, clock face, and ringer (fig. 2). The energies of significant figures from art history flow through the battery and into the transformer, marked Dada, powering the clock hands around the face, which is marked with the names

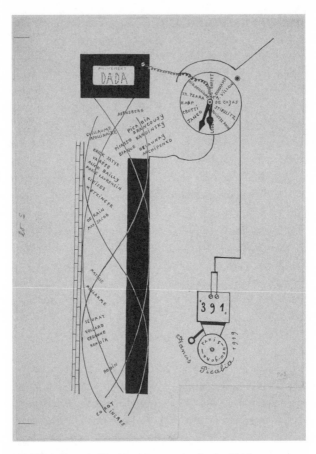

FIGURE 2 Francis Picabia, *Mouvement Dada*, 1919.

of significant Dadaists, including Arp and Tzara. Picabia himself is part of the energy inside the battery, grouped with other experimental visual artists. The bell rings out in Paris and Zurich, with Picabia's newly established little magazine, *391*, as its amplifier.

John Elderfield worries that Picabia's wiring job 'looks amateur and not entirely convincing' but makes a good case for the battery's effective functioning, noting that its positive and negative poles are carefully distinguished on the diagram.[4] However, there is another potential association with the alarm clock that Picabia's image invokes: Sarah Ganz Blythe and Edward D. Powers wonder if *Mouvement Dada* is really an image of the alarm clock's dangerous possibilities, finding in the image a 'jerry-rigged time-bomb' set to blow.[5] Both a monitoring overseer of regulated sleep and the trigger for an improvised explosive device, Picabia's alarm clock is full of contradictory and disruptive potential.

Before printing *Réveil Matin*, back when he was living in the United States, Picabia had been fascinated by mechanical drawings, especially of automobile parts. In the summer 1915 edition of the little magazine, *291*, he had drawn 'object portraits' of himself and of the magazine's editor, the photographer Alfred Stieglitz, in which they were each represented by mechanical devices. Stieglitz's portrait included (naturally) camera parts, but Picabia's was a car horn—another loud alarm intended to rouse people into alertness.[6] It seems fitting that the alarm clock replaced the car horn in Picabia's personal vocabulary after he travelled from the United States, the land of the automobile, to Switzerland, the home of clockmaking. The precision of Swiss watches and clocks was the product of a rationalising efficiency and the alarm clock symbolises the same logic when applied to the individual: it urges you to rise and grind, to optimise your daily hours, and redeem time in accordance with the Protestant work ethic. The alarm clock regulates the waking and sleeping body, readying it for work and standardising its daily rhythms to the tempo of the machine age.

Picabia's exploded and potentially explosive alarm clocks appealed to many Dadaist preoccupations. Arp, for instance, seems to have found the idea of removing a machine's cover to reveal its secrets very compelling, and described Picabia's whole period of mechanical art as an exercise in creating 'machines of the unconscious' (*Maschinen des Unbewußten*) out of hardware and machine parts, including wires, electric cylinders, wheels and screws.[7] By taking the back off his clock, Picabia was revealing the unconscious drives underneath the surface of everyday objects and, by extension, the cogs and clockwork of everyday life. For the Dadaists, who formed themselves as an avant-garde collective in response to the violence of the First World War, the appeal of a machine unconscious was a farcically irrational alternative to the enthusiasm for supposedly rational, mechanised slaughter displayed by the European ruling class.

Réveil Matin therefore records a particular set of preoccupations with, on one side, the routine of modern, mechanised life and, on the other, the desire of artists in the first decades of the twentieth century to stir people out of their half-conscious routines and conventions. In 1917, two years before Picabia printed his alarm clock, the Russian literary theorist Viktor Shklovsky would coin the term normally translated into English as 'defamilarisation', to identify what he understood to be the defining feature of art: its capacity to rouse its audience from their routinised perception of the world. In the essay in which Shklovsky defines the term, he explains that daily life makes people habituated to their circumstances and dulls their feelings and perceptions by making their responses automatic and inattentive. The automatic thinking of habit, he writes, 'eats up things, clothes, furniture, your wife and the fear of war'.[8] Art's purpose, for Shklovsky and for the many experimental writers and artists who shared this view, was to reintroduce people to seeing the world without the routine responses of habit. Defamiliarisation works by taking the familiar and making it strange, so that ordinary things can be seen again in a new way. As Shklovsky writes, art exists in order to make us 'recover the sensation of life, in order to make us feel things, in order to make the stone stony'.[9] Art, by this account, is a kind of alarm clock, rousing its audience out of habitual unconsciousness to a new alertness.

Part of my interest in alarms is, like Picabia's, a desire to get inside everyday things and see how they work. But most important to me is the exploration of how artists and writers have responded to alarms – how the annotations and signatures they make on the objects that they encounter around them can help us understand the world in a different way. Art isn't only a print taken from the texture of life itself; it's also a reworking and reimagination of the world. Art transforms, even as it represents. Picabia's alarm clock reveals more about this object than how its cogs fit together. It shows how the alarm clock has become a symbol for the dull repetition of the everyday – the same morning routine, stamped over and over again like a mechanical reproduction from a printing press – but at the same time, like the shocking art of the avant-garde, his alarm clock serves to shake us out of comfortable repose and into consciousness, to smash things apart and put them together in a different way.

*   *   *

The modern, mechanical alarm clock is not the first object to combine a mechanism for measuring time with a noise-making device. A very early example of an alarm and clock combination is a variation on the water clock (or *clepsydra*). Plato purportedly had a garden model of this device which gradually filled up with water overnight so that, by the time the bucket was full, it would overturn and spill its water and noisy pebbles to wake him and his students. Rather than having to keep an eye on the water running through (and therefore stay at least partially alert to the passing of time), the person who uses the water clock can let the alarm take over their vigilance for them, and only attend to the device when it makes a noise. By definition, then, the alarm clock is a clock that watches itself.

The modern bedside alarm clock makes a noise at a pre-arranged hour but, crucially, does not sound out any other time. An ordinary chiming clock might mark the hour or the quarter hours, but by taking away all the chimes except one, the development of the bedside alarm clock not only meant that its owner could wake up on time, but also sleep undisturbed. For an effective alarm clock, silence is just as important as the alarm. The digital alarm clock – whether a standalone device or a phone app – has a bit of an advantage over the mechanical alarm clock in this regard because the time without the alarm is truly silent, uninterrupted even by a mechanical tick.

The alarm clock wakes you up, but the real promise it makes is uninterrupted sleep, relieved of the worry that you might not wake up on time. It therefore offers a guaranteed, timed awakening, which in turn promises to help you sleep more soundly. Manufacturers of alarm clocks have naturally preferred to emphasise the restful sleep rather than the rude awakening offered by their products. An advertisement from the alarm clock manufacturer Westclox from the 1920s, for instance, describes its devices watching over their customers as 'loyal and ever-alert sentinels' (fig.3).[10]

These faithful alarm clocks are 'guardians of sleep' whose watchful attention enables untroubled rest for Westclox's customers. By alleviating the fear of oversleeping, the alarm clock enables peace of mind and better sleep – a more total unconsciousness. In accordance with the model of the alarm as a prosthetic for attention that I laid out in the introduction,

FIGURE 3 'Guardians of Sleep', advertisement for Westclox, 1929.

the 'ever-alert' alarm clock therefore takes over vigilance and allows us to lapse into peaceful sleep.

The famous opening scene of Franz Kafka's *The Metamorphosis* (1915) illustrates very well the ambivalent states of alertness that correspond to our contract with the alarm clock. At the beginning of Kafka's story, the unfortunate protagonist, Gregor Samsa, wakes up late, realises he's missed his train, and understands that he must have slept straight through his 4.00 am alarm. Among Gregor's first thoughts on waking is an indignant disappointment in his alarm clock, which seems not only to have failed to wake him but also failed to ensure him a restful night:

> Could it be that the alarm clock hadn't gone off? You could see from the bed that it was set correctly for four o'clock; it certainly had gone off, too. Yes, but was it possible to sleep quietly through ringing that made the furniture shake? Well, he certainly hadn't slept quietly, but probably all the more soundly for that.[11]

In the novel's famous first line we are told that Gregor wakes up to his metamorphosis after an unsettled night: in the original German, he awakens from *unruhigen Träumen* – restless dreams – to discover that he has been transformed into some kind of monstrous insect. The alarm clock – the *Weckuhr* or waker – has neither roused Gregor awake nor secured his peaceful sleep.

The beginning of *The Metamorphosis* is so affecting because of its combination of the deeply weird and the immediately familiar. Yes, Gregor suddenly has lots of little waving legs and a great segmented belly, but he also lies in bed fretting about whether his boss will think he is lying about being unable to work that day. The story takes the ordinary experience of waking and transforms it into something wild and nightmarish.

Having slept through the alarm and found himself transformed, Gregor is anxious that he will be visited by an insurance doctor and judged to be lazy or malingering. Kafka himself was well versed in these procedures for discerning fitness to work through his job at the Workmen's Accident Insurance Institute in Prague. His work saw him not only considering the claims of workers who had been taken ill, but also those who had been injured in workplace accidents. The Kafka scholars Stanley Corngold and Benno Wagner argue that many of his stories operate on the logic of the industrial accident, with a missed alarm or a tiny moment of inattention causing a catastrophe that must be followed by an attempt to explain the accident to a dominant institution in terms which would allow for some compensation or restitution.[12] When Gregor sleeps through his alarm, however, he's not experiencing quite the same inadvertent carelessness that might cause an industrial accident. Instead, his missed alarm is a symptom of something less eventful and more ordinary: the fatigue from the daily grind of labour, which is slowly wearing him down. This is another form of work-related injury, albeit on a more gradual timescale. Gregor's missed alarm could therefore be understood as recalling and distinguishing between two kinds of workplace hazard: the more acute dangers warned of by the industrial alarm, but also the chronic debilitation of fatigue and stress. In *The Metamorphosis*, the alarm clock doesn't warn of the slow, everyday industrial injury of overwork, but instead enables it.

Gregor's morning fantasies – sleeping late, staying at home, abandoning his job – are fantasies that we recognise even a century later. They're the dreams of unregulated sleep without an alarm, of living a life off the clock. Gregor gets to live this life in Kafka's story but it's the life of an insect, scuttling around under the furniture and horrifying his family. One way of reading *The Metamorphosis* is to understand it as a dark wish-fulfilment fantasy, in which all of Gregor's 'restless dreams' of liberation from the alarm clock come true, in a cursed and nightmarish way. And in a final, deeply ironic twist in the story's closing pages, we see his whole family eventually celebrating his death by calling in sick to their new workplaces and taking a carefree trip together to the countryside. Kafka's story recognises that Gregor Samsa has made a bargain with the alarm clock to disturb him at just the right time, in exchange for unworried rest, but it's a bargain that only needs to exist because his sleeping patterns are shaped by the unliveable demands of his work, which denies him the rest he really needs.

* * *

We have so far understood the alarm clock as a device for doing the work that Benjamin Reiss has called 'taming' disorderly sleep to fit the norms of modernity.[13] The bedside alarm clock makes sleep into the responsibility of the individual; it both meets and facilitates the demand that workers should arrive for their labour at a particular hour and is an object emblematic of a culture that valorises economies of time. When Picabia stamped his alarm clock he was re-enacting the same gesture of repetition that the alarm clock epitomises. *Réveil Matin* records the imprint of this everyday time-sense by taking the alarm clock and making the object's stamp on the paper exactly analogous to the device's repetitive mark on daily life. But we might also read Picabia's print as a sort of cathartic revenge fantasy against the tyranny of the morning alarm. Who hasn't wanted to smash their alarm clock, to refuse its call to conform to the routines of work and the demand to regulate sleep and waking according to a timetable? Rather than Gregor Samsa's doomed attempt to escape the alarm by dropping out into animality, smashing the alarm clock offers a fantasy of escape from repetitious daily time.

Harold Ramis's 1993 film, *Groundhog Day*, presents the same kind of cathartic destruction that we see in Picabia's print in its escalating violence against a hotel alarm clock, all enacted against a supernaturally heightened repetition of the everyday. In the movie, Bill Murray's grumpy weatherman, Phil Connors, wakes up to the integrated radio of the bedside Panasonic RC-6025 flip clock while reporting on

Groundhog Day in the small town of Punxsutawney. As the plot progresses and it dawns on Phil that he is living the same day on repeat, we cut from morning to morning as Phil wakes at 6am and hits the clock, each time with escalating violence. Finally, at the height of his desperation, we see the pieces smashed across the floor, as the speaker continues to play a warped and tinny version of Sonny and Cher's 'I got You, Babe'. Smashing the alarm clock offers the illusory hope of escaping from the repetitive time of the everyday, of somehow waking up once and for all.

There's something specific about the alarm clock on film which intensifies the object's association with repetition or a life stuck on repeat. When Phil wakes up on the second day in Punxsutawney, he jokes to himself that the station on the clock radio must have got the wrong tape and replayed the DJ's recorded banter from the day before. Film too can be replayed, repeated, and recut to emphasise recurrent similarities. It's a medium that has repetition built in.

A similar repetitious effect appears in the morning hours of Christian Marclay's 24-hour video installation, *The Clock* (2010), when the real-time supercut of film clocks begins to show morning alarm clocks. The alarm clocks begin to appear just before 4am, when we see Meryl Streep waking up early for work in *Silkwood*, and continue on for nearly six hours. Patterns start to appear: couples wake up together, characters successfully and unsuccessfully try and rouse each other, there are regrets about the use of the snooze function, coffee is prepared. Through Marclay's editing, the alarm clock's repetitions intensify and tropes repeat into clichés: Hugh Grant's serial oversleeping before the weddings in *Four Weddings and A Funeral*; Robin Williams's various good mornings to Vietnam. Cut together, the compressed routines of the everyday stretch out, the same again and again. If Picabia's print stopped time and exploded the alarm clock into a spatial, diagrammatic view, film's temporal medium lets us see the alarm clock's daily repetitions unfolding over time. These artworks smash open the alarm clock's secret interior, and compress time to show its temporal repetition. Ultimately, they wake us up to the alarm clock's contradictory position as both the guardian of sleep and its assailant.

*   *   *

The snooze alarm ought to reconcile some of the uncomfortable contradictions of the alarm clock that have been laid out in this chapter. It offers you, the sleeper, a choice to extend your rest and disobey the clock, to soften the shock of being alarmed into consciousness. It ought to feel like extra time, reclaimed from the tyrant clock, but surely it's time taken from yourself, either because you've disturbed your own sleep earlier than necessary, or made yourself late with supplementary snoozing? If the alarm clock is the contradictory sentinel who disturbs your rest in order to safeguard your sleep, the snooze button is the unsuccessful attempt to soften that contradiction.

When the snooze button was first marketed as an alarm clock function in the 1950s (with the General Electric 'snooz-alarm' and the Westclox 'drowse' button), advertising presented snoozing as a luxury. It was imagined to be a distinct category of extra and indulgent sleep, for pleasure rather than for rest. Moreover, it was a period of sleep that you could buy through the purchase of a particular item. As Jonathan Crary has argued in *24/7*, his book about the encroachment of non-stop capital onto all areas of waking life, 'Sleep poses the idea of a human need and interval of time that cannot be colonized and harnessed to a massive engine of profitability, and thus remains an incongruous anomaly and site of crisis in the global present'.[14] The advertised function of the snooze alarm was correct, of course: if we live by the logic that every waking moment must have its value extracted from it, then time spent sleeping *is* a luxury – you snooze, you lose. But the snooze button also represents an attempt to divide sleep into pieces, to colonise portions of it and to harness them for profit. The time of snoozing or dozing is the border zone between sleep proper and waking, and it was therefore available to capital as a luxury that could be purchased with the function of the snooze alarm.

The snooze button therefore represents bonus time that you've stolen from yourself. It entrenches the already weird temporality of the alarm clock, which holds the self of the morning accountable to the self of the night before. As the device through which we structure our sleep and waking to fit the demands of work, regulating ourselves to the rhythms of capital, the alarm clock is the model for the many modern devices discussed in Chapter Six which promote wellness and productivity through self-regulating practices of monitoring.

The artists and writers of the modernist avant-garde imagined that we could use art to wake ourselves up to a new understanding of the world. But the alarm clock more often seems to hold us to the commitments of the night before ('I'll wake up early tomorrow and finish writing this chapter about the alarm clock . . .'). In doing so, it turns an opportunity for a transformative awakening into just another continuation of the same.

2 FIRE

'The smoke detector is an inert item – not like a coffee pot which makes something every morning': this was the case made by an executive from the North American arm of the Philips corporation, in 1977, for pouring money into advertising their new line of smoke alarms.[1] In the late seventies the domestic smoke detector was still a relatively new technology, and there had been a rapid expansion in the market as consumers were more and more convinced of the device's reliability. Nevertheless, smoke alarms would not sell themselves, the ad-man argued, because there was no tangible and immediate consumer benefit to owning one. Unlike a coffee pot, which gives a daily reminder of its utility, a smoke alarm's ideal state is inert: customers buy it and hope they will never have to use it.

The domestic smoke alarm is one of the most important safety interventions of the last century. However, it is easy to overlook precisely because of its unobtrusiveness. As I've been arguing, the alarm's job is to relieve us of the burden of constant vigilance. The inertness of the smoke alarm – its minimal demand on attention when all is well – is a heightened version of the promise offered by every alarm to watch and to warn, to act as a guard or sentinel. In its inertness, the fire or smoke alarm is an object that activates the people who use it. Around it, safety practices such as drills, evacuation plans, simulations and other preparedness activities circulate, all under the watch of the inactive alarm.

The Statitrol corporation, the firm that marketed the first battery-operated domestic smoke detector, has a fascinating archive, which has been completely digitised by Worcester Polytechnic Institute in Massachusetts.[2] The papers of Duane Pearsall, the founder of Statitrol and developer of the technology behind the device, record how a set of cultural, social and political practices developed around the smoke alarm.

Statitrol began as a firm that developed and manufactured devices that were used by photo labs and printing operations to control static electricity. In 1965, workers in one of the company's research labs were attempting to interpret the readings from a new version of their static neutraliser. To test whether it was working properly they had put together a makeshift ionisation detector so they could measure the output of the static neutraliser's ion generator. Duane Pearsall describes in his memoir how an ion meter was 'kludged together' by one of the company engineers.[3] This makeshift instrument seemed to be returning wildly unexpected readings, however, and attention began to turn to one of the lab assistants who had a habit of chain-smoking in the lab. It became clear that there was a correlation between the lab assistant blowing smoke from his cigarette and the readings dropping to zero. The particles in the smoke were absorbing the ions and stopping the electrical current, breaking the circuit and sending the meter down to zero and then back up again as the smoke dissipated.

Earlier fire alarms had used heat detection or photo-optical sensors to detect smoke, but these methods were a bit more expensive and had a higher rate of false alarms. An ionisation smoke detector (which used a small amount of radioactive material to generate the ions, which were then detected by the device's ion meter) promised to offer high sensitivity and therefore a longer evacuation time for homes where people might need to be roused from sleep in the case of a fire. The team decided to replace their instrument's measuring dial with an annunciator, turning their meter into an alarm.

The domestic smoke alarm that Statitrol would go on to develop, the SmokeGard 700, was distinctive not just in its use of ionisation for smoke detection, but also in the way it addressed its target market. The device was marketed direct to consumers through the Sears Roebuck catalogue, beginning in 1972. The imagery that's familiar to us from alarm clock advertising also appeared in the early product information for the SmokeGard, establishing the smoke alarm as a replacement for one's own vigilance. For instance, early marketing materials for the SmokeGard 700 and 800 used the tagline, 'You'll sleep better knowing your SmokeGard Smoke Alarm never does!' (fig. 4).[4] The same document also

You'll Sleep Better Knowing Your SmokeGard® Smoke Alarm Never Does!

SmokeGard®

MODEL 800A EARLY WARNING HOME SMOKE ALARM

FEATURES OF THE SMOKEGARD LOW COST IONIZATION SMOKE ALARM —

- Designed for *Life Safety* — Alerts the family before smoke or toxic gases accumulate
- Ionization Principle — A new, highly sensitive electronic home smoke alarm
- Detects smoldering fires earlier — yet will not alarm in a room full of heavy smokers
- Battery powered — no wiring — fastens to ceiling in minutes, with screws provided
- Readily-obtainable 1.5 volt AA alkaline batteries
- Fail-Safe — batteries last 12–16 months — horn will "click" when the batteries are weak Exclusive removable mounting plate

- Unique pilot light pulses every 5-10 seconds — assures you that batteries are capable of sounding an alarm
- One alarm protects each bedroom area
- Location — on the ceiling of the hall outside bedroom doors (approved for both wall and ceiling mounting)
- Alarm horn — 85 decibels — loud enough to awaken most heavy sleepers through closed bedroom doors
- Suitable for homes, mobile homes, apartments, vacation trailers or cabins
- Extra units can guard basement and living room
- Underwriters' Laboratories and I.C.B.O. Listed

FIGURE 4 'You'll Sleep Better', product information for the SmokeGard, 1972.

emphasises the battery option for the device, which meant it could be fitted by even the most amateur DIY-er in some unobtrusive place in the house, not constrained by mains power access. Running on battery power created the need for further reassurances about reliability, however, which was achieved by providing a pilot light and a second alarm to indicate a low battery (more on this later). More recently, both battery power and ionisation alarms have begun to be phased out and even legislated against in some jurisdictions, but both of these elements of the SmokeGard's design were a revolution in domestic fire safety in the 1970s.

Because it doesn't require an electrician to fit or service it, the battery-operated smoke alarm places installation and maintenance responsibilities wholly onto the person who purchases the device. Looking through the archive of marketing and instruction materials for the SmokeGard, it's clear that a specific kind of person is imagined as the consumer. Their hand appears fitting the alarm; there are images of their sleeping children; there are references to their basement and their vacation home: this is a consumer who is conscientious and self-sufficient, who has responsibility for the space in which they live, and therefore responsibility for managing the risks and hazards within it.

In both the marketing information for the SmokeGard and in fire safety information of the same era from the US National Fire Protection Association, the category of the 'household' emerges as the unit protected by these domestic fire alarms. The diagrams in these documents reveal some

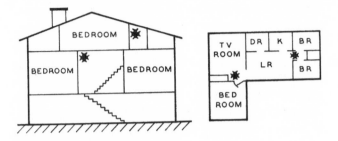

FIGURE 5 Smoke detector placement diagram from the National Fire Protection Association, 1975.

assumptions about norms for where and how people live (fig. 5).[5] These imaginary households are single-family homes (not apartments or multiple-occupancy units with shared hallways or living spaces); they're detached from each other, apparently without shared walls or floors or roofs.

In these diagrams, the smoke alarm creates a zone of protection that's exactly the same size as the household which, in turn, is imagined as a nuclear family unit. The head of the household is responsible for the safety of the occupants within the home. It's a way of thinking about fire risk as something which can be controlled by the responsible decisions of an individual consumer, and which is atomised into units of protection that map onto the individual home and the nuclear family.

This individualising approach to fire risk was echoed very strongly in political discourse from United States in the 70s. Investigations from the early 1970s by the Presidential

Commission on Fire Prevention and Control were prompted by data that reported that the country led 'all the major industrialized countries' in the number of per capita deaths and property losses from fire.[6] The Commission's 1973 report, *America Burning*, emphasised individual citizens' responsibility to prevent fires, with laments about their 'inattention', 'carelessness' and 'indifference' appearing throughout the text. President Nixon's opening comment at the beginning of the report is instructive:

> Only people can prevent fires. We must become constantly alert to the threat of fires to ourselves, our children, and our homes. Fire is almost always the result of human carelessness. Each one of us must become aware – not for a single time, but for all the year – of what he or she can do to prevent fires.

Given that constant attention to the threat of fires was an impossibility, and the fact that the report eschewed big government interventions in fire safety, the prosthetic vigilance of a domestic smoke alarm offered a market solution to this public safety problem.

In the summer that I was writing this book, my next door neighbours in inner-city Liverpool had a kitchen fire. After the firefighters had extinguished the fire, they went up my street with a box of smoke alarms and a step-ladder, fitting extra alarms for anyone who wanted one, from my elderly neighbours to the students renting multi-occupancy homes. Our Edwardian terraced houses are small and closely packed, with shared walls and roofs, and smoke from even a small housefire travels across the back alley to the next row of houses. The local fire brigade told us that collective fire prevention keeps the whole street safe.

While the mid-century battery-operated smoke detector would undoubtedly save many lives, it also represents the safety culture of its time, which placed responsibility on individuals to be, as Nixon put it, 'constantly alert' to danger. Rather than conceiving of fire safety as a collective endeavour, the domestic smoke alarm-extended monitoring vigilance only up to the limit of the family unit and of the domestic household, and no further. Ultimately, the smoke alarm circulates within a safety culture that is shaped by broader political currents that tend to make individuals completely responsible for their own safety and welfare, and which emphasise the family and the household as the basic units of social life. The domestic smoke alarm is therefore not an ideologically neutral object. We are now ready to encounter some representations of the domestic smoke alarm that are explicitly about the alarm's capacity to organise safety cultures and to generate practices for managing risk and danger.

*   *   *

As we have already observed, the direct-to-consumer marketing and the ease of its home installation meant that the smoke alarm in the 1970s United States affirmed an understanding of private domestic space, the family unit, and household security that went hand-in-hand with the norms of suburban home-ownership. Don DeLillo's 1985 novel *White Noise* is well-known for its depictions of the life of a household under conditions of risk, threat, and disaster, and is a good illustration of how the battery-operated domestic smoke detector participated in the broader safety culture of the period.

The novel follows Jack Gladney, a college professor in a mid-western town who narrates his worries about death and disease and his ever-present anxieties about his family's safety. In one of the earliest scenes of the novel we are introduced to Jack's large family, who are eating lunch in their kitchen. They all make sandwiches, Jack's wife regrets her food choices, and the scene ends with a flat and darkly funny observation from Jack: 'The smoke alarm went off in the hallway upstairs, either to let us know the battery had just died or because the house was on fire. We finished our lunch in silence'.[7]

The novel's title, *White Noise*, invites one potential interpretation of this scene. The Gladney house is full of domestic white noise: the radiators chirp, the washer and dryer vibrate, the thermostat buzzes, the refrigerator throbs, the waste-disposal unit grinds. Against all of this background noise, who could pick out the smoke alarm as being particularly worthy of attention? The scene conveys something of the Gladney family's struggle to identify those messages that are important within a culture full of mechanical sound. Alarms, even when they are activated, are easy to ignore in an automated home full of white goods making white noise.

Thus, if the Gladney house conjures up a dream of home automation, it's a dream that has soured. Domestic appliances promised that activities in the home which had previously required time, labour and, crucially, vigilance could now take place with minimal user involvement. Think of the difference between watching over a pan of soup on the hob vs popping it in the microwave until it pings; the ping is what calls you back, but until that point your attention can be elsewhere. In the Gladney house, these consumer appliances have become oppressive and demanding. Later in the novel we see Jack hauling out the family's discarded (and finally silent) consumer items – 'Plastic electric fans, burnt out toasters'– and depositing them for waste collection, 'waiting for a sense of ease and peace to settle in the air'.[8] Rather than alleviating the responsibility for vigilant alertness with their watchful and inert presence, these machines disturb the peace.

We can therefore think of the smoke alarm as falling both inside and outside of this category of consumer objects that filled a modern home with beeps and pings and buzzers and chimes. The marketing executive from the beginning of this chapter faced a challenge in how to advertise the smoke alarm, a device which was both like a toaster or an electric coffee pot, but at the same time very unlike them. The smoke alarm belongs to that category of household alerting devices that turns their owners into overseeing managers of their self-monitoring appliances but, unlike the consumer luxuries of the tumble dryer or the microwave (or, slightly earlier, the bedside alarm clock), the smoke alarm also has to present itself as the household's defence against danger. It's the smoke alarm's association with risk or hazard that makes it particularly important for Jack Gladney as the patriarchal head of his household.

In a time period when home automation was more often associated with savings in feminised domestic labour such as cooking and cleaning, the smoke alarm is an unusual household device that is associated with the automation of the father's domestic role of vigilant protector. The first of many occasions on which we see Jack's anxieties in monitoring his family's safety appears in the novel's opening scene when he looks at other, wealthier fathers and thinks that something in their bearing speaks of 'massive insurance coverage'.[9] DeLillo's own father was a clerk at the Metropolitan Life Insurance company, a detail that adds weight to the association between an anxious, responsible patriarchal masculinity and the calculations of risk and hazard that underpin actuarial calculations.[10] DeLillo's novel suggests that the safety culture associated with the domestic smoke alarm places a weighty burden on the head of that household to anticipate and manage its threats.

* * *

The US post-war consumer boom was, of course, underpinned by the geopolitics of the atom bomb, another inert technology which, like the smoke alarm, offered a guarantee of security by its non-activation. Perhaps it's too far-fetched to imagine the little grain of nuclear material in the household smoke alarm as a cryptic reminder, at the heart of the home, of the bigger fears of nuclear threat.[11] But if the management of nuclear risk meant creating an invisible system of alerts and sirens – inert technologies that it was hoped would never see use – the smoke alarm was certainly the home's equivalent of the nuclear warning system. It's therefore possible, in this context, to tell another story about the mid-century invention of the smoke alarm that could place it within a broader transformation of the concepts of risk assessment and risk management.

By the second act of DeLillo's *White Noise*, the threat of a housefire has returned on a much larger scale, as a looming catastrophe which threatens the whole town. The disaster is named non-specifically as the Airborne Toxic Event and described in the novel as a 'dark black breathing thing of smoke', inviting an association with the Gladneys' smoke detector.[12] In contrast to the opening scene with the disregarded smoke alarm, the community soon begins evacuation proceedings accompanied by the 'noise of alarms from all sides': 'sirens from ambulettes', 'a clanging bell', and 'a series of automobile horns' rising to 'a herd-panic of terrible wailing proportions'.[13] Awakened from inertness, alarms and sirens are suddenly everywhere in the community beyond the Gladney house.

DeLillo's Airborne Toxic Event has often been read in dialogue with ideas about risk and accidents that were emerging at the same time as *White Noise* in the 1980s. The work of the sociologist Ulrich Beck, for example – which argued that his society was undertaking a huge project characterised by attempts to try and manage the risks caused by human activity – speaks to many of the concerns about safety that DeLillo reconfigures down to the scale of domestic drama. Beck's 1986 book *Risk Society* argues that modern society has to grapple with the understanding that its risks come primarily from human actions (such as nuclear weapons or industrial accidents, climate breakdown or pandemic diseases) and that these risks have increased dramatically in scale and effect. Rather than abandoning the risky business of nuclear power or industrial-scale animal agriculture, science and industry try to manage risk, to find its 'tolerable' limits.[14] In *White Noise* we can see the different faces of what Beck called the 'risk society'. The Airborne Toxic Event, for instance, demonstrates the way that risk becomes free-floating, drifting cloudlike away from a point of origin. Beck's aphorism, 'poverty is hierarchic, smog is democratic', registers how the hazardous consequences of wealth-generating technological advances trickle down or float around, while the wealth itself tends to concentrate in one place.[15] We could therefore understand DeLillo's Airborne Toxic Event to be a floating reminder of the distributed hazards of industrial modernity. The way that a risk society prefers to deal with these dangerous consequences is to insist that individuals ought to protect themselves and their households against hazards by purchasing such protections as smoke alarms or life insurance. It's therefore possible to understand Jack's realisation that he does not have the same 'massive insurance coverage' as other parents as the very beginning of a recognition that risk has been individualised and is unevenly, and therefore unjustly, distributed.

With modernity's new attitude to risk comes a set of activities for assessing and managing hazards through practices such as risk assessments, drills, disaster preparedness strategies, and emergency simulations. All of these practices involve forms of anticipation which imagine the present as a kind of rehearsal for future dangers. In *White Noise,* the scene with the Gladney family's smoke alarm reads, in retrospect, as a failed drill or preparedness simulation for the emergency of the Airborne Toxic Event. But this prefiguring is just the start of all sorts of strange echoes and repetitions which emerge whenever drills and simulations are mentioned in the novel. The evacuation of the town is overseen by an organisation called SIMUVAC, whose purpose is to set up simulated evacuations as a way of planning for real disasters. But the Airborne Toxic Event is, apparently, a real disaster. Simulation and reality have somehow got mixed up together, and it's not possible to tell real alarms from false, event from drill. The Airborne Toxic Event evacuation is an instance where the real disaster has been mistaken for a drill – the rehearsal replaced with a performance. And reading back to the Gladneys' kitchen, we encounter a kind of Beckettian drama where, in spite of all the scripting and rehearsal that ought to preface the alarm's direction to the family, they do not move.

One of the things the example of the Gladneys' smoke alarm tells us about alarms, then, is that they mediate between two extremes of under-reaction and over-reaction. Alarms should *prevent* alarm because they should fit into a bigger safety structure of instructions, evacuation plans and risk assessments, so that people keep calm and know the drill when the alarm goes off. But they are also there to alarm us out of our own under-reactive inertness when an emergency demands it. This mediation between particular affects – the allocation of calm and alarm – is at the core of the alarm's more far-reaching administration of feelings.

*   *   *

Fortunately, the Gladneys' smoke alarm was not alerting them to a housefire, but instead to the fact that the alarm's battery had just died. To consider the significance of this detail, we should come back once again to the specifics of the Statitrol SmokeGard. Its power source, using a battery rather than being wired into the mains, resulted in another technological development which went hand-in-hand with the product's battery-powered convenience: the necessity for a secondary alarm to alert users when the batteries began to run down. The SmokeGard's 'fail-safe' was an important reassurance and reminder of the activity of this otherwise inert technology. It was such a distinctive part of the product

FIGURE 6 'CLICK', advertisement for the SmokeGard, 1975.

that the company's battery-replacement service built an arresting publicity campaign around its deliberately annoying low-battery warning sound, advertising the 'CLICK, CLICK CLICK, CLICK CLICK CLICK CLICK' of the fail-safe as one of its key safety features (fig. 6).[16]

The click of the low-battery alarm punctuates the alarm's inertness. It serves as a reminder that the alarm is present and doing its job, even when it isn't activated. A low-battery signal can also operate as an opportune reminder of the safety practices that need to be rehearsed and drilled to prepare for the alarm's activation.

The first play of Michael Frayn's sequence of very silly technological farces, *Alarms and Excursions: More Plays than One* (1998), uses some of these ideas about the alarm fail-safe and alarm preparedness to stage a comedy about alarms and our response to them. The opening piece of the sequence, 'Alarms', is set at a dinner party where two couples come together for a quiet dinner. Their evening is soon interrupted by something mysterious that goes 'chink'. The characters attempt to hunt down the source of this warning sound (rooting out a pager in a pocket, a malfunctioning electric corkscrew, and a persistently buzzing oven timer) until the situation escalates and the whole house is alarmed with a networked system of phone ringers in every room, a shrieking burglar alarm, and a smoke detector that may or may not have run down its battery. Into all this, a threatening voice speaks to John, the host of the dinner party, leaving an extended message on his answer machine (after yet another beep) to urge him to take action over some unspecified responsibility he has been attempting to ignore for years. By the end of the play John is locked in his empty house, ignoring a flood of alarms. The final stage directions read,

> *Doorbell. Door-knocker. Hammering on door with fists. Flapping of letter box. Buzzer. Car-alarm. Police siren. The burglar alarm goes off.*[17]

As the curtain falls, the voice speaking to John through the answerphone seems to be about to tell him what to do to get himself out of this mess – almost like a stage director instructing a waiting actor – but, as with the Gladneys, we don't see his response.

Frayn's play reserves special attention for the household smoke alarm, and especially for its fail-safe. Nancy (the group's Cassandra) warns that the noise they are hearing may be the low-battery alarm:

Nancy I think it may just need . . .
Nicholas Yes, we heard you.
Nancy . . . a new battery.
Nicholas Don't be silly. Things don't go . . .
    Chink.
    . . . to tell you they need a new battery.
    John *reads the instructions.*
John It needs a new battery.[18]

Another character gives the best, most succinct description of this feature, as he speculates about whether the sound is therefore 'not a smoke alarm' but 'an alarm alarm'.[19] As an 'alarm alarm' the fail-safe represents absurd alarm proliferation – the possibility of alarming stretching on infinitely, with an alarm on the alarm that's on the alarm alarm, etc. The proliferation of alarms becomes overwhelming and threatens the peaceful inattention that the inert alarm promised in the first place.

As in *White Noise*, the characters in Frayn's 'Alarms' live in a noisy home full of demanding household appliances. In performance, the buzzing and chinking of these alarms must come in with perfect comic timing to complement the dialogue:

Nicholas It's the oven.
Jocasta The oven?
Nancy You said you'd got something in the oven.
Jocasta The oven doesn't go . . .
    Chink.
Nancy No, but the timer thing.
Jocasta The timer thing? The timer thing goes . . .
    Buzz.
    Excuse me . . .
    She goes off right.
Nancy *Something* in the kitchen, though. The microwave.
Nicholas Or the food-processor.
Nicholas The coffee-grinder.

The buzzing stops.
Nicholas The toaster.
Jocasta The toaster?
John How can a toaster go . . .?
Chink.[20]

In contrast with real alarms that interrupt and disrupt ordinary routines, these alarms are orchestrated and cued to fit into the performance, to become punctuation or a punchline. A gap therefore opens up between these theatrical false alarms and real alarms, which must be intrusive and disruptive (because they're unexpected) rather than perfectly timed.

Earlier in the play, John's wife, Jocasta, has suggested that they simply ignore the mysterious chinking sound and carry on with their meal, but he gets increasingly frustrated, saying, 'But it's plainly trying to tell us something. . . . But it's plainly *meaningful!*'[21] John's exasperation makes me wonder just how *plainly* an alarm really communicates. The problem that the characters have in the play is that without finding the source of the 'chink' they can't interpret its meaning – its meaning is not plain at all. In DeLillo's *White Noise*, when Jack hears the smoke alarm and isn't sure if the batteries are dead or if the house is on fire, this is because the alarm can only draw attention to itself with its noises; it can't give further information to distinguish between 'replace my battery' and 'evacuate the premises; the house is on fire'. Presumably one of the Gladneys must, eventually, go upstairs and investigate, but in the novel we're never told what the alarm was saying. More recent smoke detectors, which vocalise as well as beep, are able to do exactly that, and can report directly that the battery needs replacing. Nevertheless, all of the sounds in Frayn's play, from the chink and the sirens to the answerphone voice, are important because they know something the characters don't. The alarms exist in the first place because they need to alert the household to something that its inhabitants haven't been monitoring: as one of the characters says, 'A smoke alarm must know more about smoke than you do!'[22] The alarm has been watching the things that we can't, extending our alertness with its prosthetic vigilance.

As with Frayn's most famous farce about farces, *Noises Off* (1982), *Alarms and Excursions* takes a stage direction as its title, and has a similarly metatheatrical element to the attention it draws to its own noises off, in the form of these orchestrated alarms. The stage direction 'alarms and excursions' itself was originally an instruction to stage something that would give the general impression of busy (usually military) activity, of springing into action by minor attendants on the early modern stage. It's, in essence, a stage direction to tell actors to depict an alarm and the response to it. The instruction is interesting as a short-hand, non-specific stage direction which allows for improvisation and unscripted physical movement. Idiomatically, 'alarms and excursions' has also come to describe gestures or activities that give the impression of busy movement. That these activities might be merely performances of purposeful action also seems to capture something of the theatricality of *any* response to an alarm. The flurry of 'excursions' that ought to follow an alarm are somewhere oddly between the rehearsed and the improvisatory. When an alarm is raised, the actions that follow are inevitably confused and chaotic, but the response to an alarm should also be organised and practised.

Metatheatrical plays like *Noises Off* and *Alarms and Excursions* often remind us that drama is something that's rehearsed and repeated as well as spontaneous and extemporised, and that what's interesting about theatre is the mixture of the two. A stage direction like 'alarms and excursions' records exactly that combination of direction and improvisation. In Frayn's 'Alarms', this dynamic between chaotic, farcical improvisation and scripted preparedness is clearly displayed in the characters' response to the mysterious 'chink'. After trying to hunt down the source of the noise, their next recourse is to search out the kitchen drawer full of instruction leaflets for the household's gadgets, like four characters in search of a script. Like an under-rehearsed cast, they are improvising their way through a situation that should have been choreographed and cued. They are trying to get out of the improvisatory and into the rehearsed, and attempting to respond to the alarm in a way that (for safety's sake) ought to have been prepared in advance.

* * *

In a speech to the World Economic Forum in January 2019, Greta Thunberg used the metaphor of a house fire to talk about the climate crisis: 'Our house is on fire,' she said, 'I am here to tell you, our house is on fire'.[23] We've seen how the battery-operated domestic smoke alarm reinforced a safety culture of individual responsibility and reflected a broader understanding of risk that saw the hazards of industrial modernity – represented by smoke or airborne toxins and smog – spread out over populations who often reaped little of the wealth from the industries that generated them. The same discourses are in place today to individualise mitigations against climate change, with continued entreaties to consumers and households to focus on their carbon footprints and consumption habits, rather than enact shared, structural solutions. The way that the mid-century household was conceptualised as the limit of risk and responsibility for fire safety has, again, been echoed in the lines that are being redrawn around nations, as world leaders attempt to shift responsibility for carbon emissions to other countries, indulging in the fantasy that risk can be contained and managed within the confines of a single location. Thunberg's metaphor of the house of fire reversed the tendency of safety cultures to entrench individual responsibility and to imagine risk mitigation that, at most, stretches to the bounds of the household. In addressing world leaders with an alarm call about a housefire, she demanded that they understand the whole planet, unbounded, as the house. It's a house that we share, and in which actions in one place affect life in another.

But discourses of alarm related to climate politics do not always work in quite the same way. Declarations of climate emergency and the slogans of climate activists proclaiming, 'This is not a drill', are an attempt to sound an alarm call to attention and to action.[24] In doing so they activate exactly the cultures of rehearsal, risk-management and preparedness that we recognise from the more ordinary practices of fire safety. However, there isn't a clear path for ordinary people who hear these alarms to take to produce a response that's politically effective. Like the characters in Frayn's play, we're under-rehearsed and unprepared; we haven't been through a drill before the alarm has gone off. Unlike world leaders, individuals are not able to make decisions that will protect their shared home from catastrophe. Without a political object for attention-grabbing, consciousness-raising activities, a good-faith declaration of emergency can leave us in the position of Cassandra, watching powerless as the consequences of climate catastrophe are unevenly distributed around us.

In an era of climate catastrophe, we need an effective understanding of the principles of alarm response more than ever. But if our reference points continue to be shaped by the safety cultures of the twentieth century, there's a danger that we will continue to countenance either limited, individual, household, or national, solutions to collective emergencies, or to understand awareness of a problem as being an end in itself. It's only by recognising the history of alarms and the contingency of the safety cultures that come along with them that we can formulate effective strategies for calling people to attention and to action.

3 SECURITY

There's a man skipping, spinning, flipping and flexing his way through a thicket of blue laser beams. The soundtrack is a cool Moroccan-French hip hop instrumental. The setting is an art museum, with the black-clad man's dynamic shape flanked by rows of white marble statues. He's very nimble. He probably knows capoeira. This is the laser dance scene from *Ocean's Twelve* (2004), where the Night Fox (a master thief, played by Vincent Cassel) recounts how he made his way through this laser maze without triggering the security alarms. It's a set-piece that makes a spectacle out of virtuoso alarm avoidance.

In the heist or caper we find a whole genre which, like this scene, takes its shape from the skilful evasion of alarmed security measures. It's a genre which therefore certainly deserves our attention in a book about what artistic and cultural objects can tell us about alarms. It's generally accepted that the heist film emerges as a distinct genre with two landmark films: *The Asphalt Jungle* and *Armored Car Robbery*, both released in 1950. Both films have genre-defining moments involving alarms.

Armored Car Robbery starts with its master gangster using a stopwatch to time the police response to a faked alarm call, and therefore figuring out the three-minute window of opportunity available to his crew for their robbery. In *The Asphalt Jungle* we see one of the crew physically negotiating the 'electric eye' of the jewellery store's security sensor by inching, supine, along the ground. The alarm therefore establishes constraints of time and space that the thieves must work around. *The Asphalt Jungle* also uses the sounding of security alarms to underscore the personal qualities of planning and patience of its master thief, Doc Riedenschneider. In one memorable scene, Riedenschneider smokes a cigar, unbothered, after one of the thieves has set off an entire block of alarms blowing up the safe with his home-brewed explosive. From the outset, then, the alarm sets the terms for some of the key features of the heist: first, the constraints or challenges that are overcome by planning and skill and, second, the emotional control of those involved in the crime.

I've been arguing so far in this book that the alarm is a device for organising feelings – for toggling between inertness and alertness, for managing fear and anxiety, for making space for sleep and safety. But fictional thieves like the Night Fox and Doc Riedenschneider do not use alarms correctly. Remaining suavely unalarmed, the thief refuses to feel the feelings the alarm administers. Not only does he remain resistant to the alarm's effects, but he also manages to operate exactly within the places the alarm does not reach.

I want, then, to consider how the security alarm offers us a spatial understanding for alarms and for attention beyond or outside the alarm's zone of vigilance. What varieties of playful, aesthetic, or unauthorised attention operate beyond the alarm's watch? And what does our fascination with heists and capers, with characters who specialise in the masterful skirting of security, tell us about the relationship between art and alarms?

* * *

The 1853 patent for the first electric security alarm describes it as a 'magnetic alarm' which can be applied to the doors or windows of a house 'for the purpose of giving alarm in case of burglarious or other attempts to enter the same'.[1] Using the electro-magnetic contact switch technology developed for the telegraph, the device's inventor, Augustus Pope, had found a way of creating an alarmed perimeter around buildings (fig. 7). On the face of it, Pope's invention intensified the secure dividing line between interior and exterior, private and public space. Unlike a mechanical chime or bell that could be tripped on entry, Pope's burglar alarm could become part of the fabric of the building, with wires hidden under wallpaper or in grooves in the floorboards below the carpet.

The older protections of walls and locks now had a new piece of security architecture. But just a few years later another inventor, Edwin Holmes, bought the patent from Pope and developed both the technology and its marketing

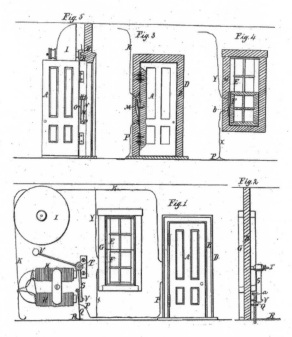

FIGURE 7 Diagram from Augustus Pope's patent for an electromagnetic burglar alarm, 1853.

in ways that challenged the simple model of perimeter security that was implied in Pope's invention.[2]

One of the first things which Holmes did was to develop a sophisticated marketing strategy for Pope's alarm. Among his sales tactics was the construction of a little model building with a tiny hinged door and window and an outsize bell on the top to sound the alarm, intended to demonstrate the workings of the device to prospective customers. Holmes's son reproduced a photograph of the model in his 1917 book about his father's life.[3] Customers could play at being the gigantic burglar trying to break and enter through the six-inch door and could see the bell ring for themselves.

Holmes's other important development was to take the burglar alarm beyond an isolated perimeter system. He was able to hook individual alarms to the telephone network and therefore implement a central station to which alarms could be routed. Rather than reinforcing self-contained, individual homes and businesses with an alarmed perimeter, Holmes's system ultimately connected them together, blurring the inside and the outside, the private and the public. The private space of the home or the safe protection of the bank vault were understood to be most secure when they could be monitored from somewhere outside.

Burglary is a specifically spatial crime that relies on codifying a legal concept of inside and outside. As distinct from more general acts of theft, burglary needs architecture, needs the boundary lines of a structure. The crime of burglary specifically occurs *in here* rather than *out there*.

The burglar alarm alerts us to the boundary between inside and outside, a division which has a legal as well as a practical status. The sound theorist Brandon LaBelle has observed that alarm systems therefore serve to puncture closed spaces, even while they seem to be preserving their security intact:

> Placed upon the home, alarm systems weave together public alert with the security of private life. . . . [T]he home is perforated by the circulation of various signals and media, networking the domestic into a larger infrastructure of communication, entertainment, and security that cut into the home while monitoring such perforations.[4]

Security systems both pre-empt and repeat the act of burglary by boring a hole through seemingly enclosed and private space, rupturing its surfaces and opening it up to a world outside. This account of the security alarm resonates, I think, with some of the discussions of the smoke alarm in the previous chapter, in which we saw how the domestic smoke alarm narrowed the perimeter for safety and risk-management down to the boundaries of a building. Burglar alarms, historically, have done the opposite, by alerting those *outside* the building to a forced entry, cutting through the very walls that define the burgle-able space.

*　*　*

Cultural representations of burglars and burglary have been fascinated by this spatial property of the security alarm which both maintains and undermines the boundaries of property and privacy. David Foster Wallace's 1996 novel, *Infinite Jest*, features as one of its main characters a burglar named Don Gately who has a specialised method for disabling security alarms. An early scene describes him breaking into the home of an Assistant District Attorney to exact revenge for some previous charges and details his signature technique at length:

> Gately and the associate went that night to the A.D.A.'s private home in the upscale Wonderland Valley section of Revere, killed the power to the home with a straight shunt in the meter's inflow, then clipped just the ground wire on the home's pricey HBT alarm, so that the alarm'd sound after ten or so minutes and create the impression that the perps had somehow bungled the alarm and been scared off in the middle of the act.[5]

Gately and his associate pretend to have been alarmed by the alarm and give the impression they abandoned the scene. In fact, Gately has planned his revenge carefully and spent his time in the house photographing his accomplice and himself doing something unspeakable with the household toothbrushes, Polaroids of which he will send to the address a couple of weeks later. Like the Night Fox or like Doc Riedenschneider, Gately is not alarmed by the alarm, and doesn't submit to its management of affect or attention.

The 'straight shunt in the meter's inflow' is Gately's modus operandi. The same terminology appears again in a slightly later description of another of Gately's break-ins, this time of a home with a 'chintzy SentryCo alarm system that fed, idiotically enough, on a whole separate 330 v AC 90 Hz cable with its own meter'.[6] This meter too is 'straight-shunted' by Gately. However, this burglary goes horribly wrong, a man dies, and Gately is identified precisely because of his 'signature M.O.' of 'killing power with straight shunts to a meter's inflow'.[7]

To investigate what this 'straight shunt' method might mean and where it might have come from, I looked to Wallace's own book collection. The Harry Ransom Center catalogued his books after his death in 2008, and their holdings list a book by Wayne Yeager called *The Techniques of Burglar Alarm Bypassing* as part of Wallace's library. However, none of the language that recurs in the descriptions of the alarm tampering in *Infinite Jest* (the meter, the inflow, the straight shunt) is present in Yeager's book. There is something curious about the idiosyncratic terminology that Wallace chose, seemingly very deliberately, for his master alarm-bypasser. Later in the novel, Gately recalls that he was informed of his mother's illness with some very specific and similar terminology:

No one tells you when they tell you you have cirrhosis that eventually you'll all of a sudden start choking on your own blood. This is called a *cirrhotic hemorrhage*. Your liver won't process any more of your blood and it quote *shunts* the blood and it goes up your throat in a high-pressure jet, is what they told him.[8]

The word 'shunts' – so new to him that he mentally quotes it and groups it with the medical terminology of the cirrhotic haemorrhage – is the word that he will subsequently come to use to think of shutting off burglar alarms. Towards the end of the novel we see a final use of the word 'shunt', again in a medical context, after Gately has been shot in the shoulder. In the hospital, worrying about whether his voice might be affected by his injury, he imagines his shoulder and voice box linked by 'fucking shunts and crazy interconnections with nerves'.[9] It's as though the electrical nerve impulses of his own body have been clipped, shut off and rerouted like a compromised burglar alarm, resulting in his fear that his mind has been somehow disconnected from his voice box by the bullet.

Tracking the word 'shunt' from its association with Gately's alarm-shunting technique to his fears about bodily shutdown, we can see that there's a relationship between the thwarted alarm system – its shunted inflow, its clipped wires – and the body's dysfunctions. Moreover, in both Gately's case and his mother's, it's the throat, the mouth and the voice that are affected, with these shunted nerves and blood flows resulting in compromised speech. The throat therefore becomes the equivalent of an alarm's annunciator, silenced by the bypassing shunt to the system.

A shunt is a diversion, a rerouting, a bypass, so Gately's 'straight shunt' seems slightly oxymoronic from the outset. But thinking about burgling from a spatial perspective reveals some ways of reconsidering this imagery of direct routes and diverted paths. Geoff Manaugh, in his fascinating book *A Burglar's Guide to the City*, makes the argument that the burglar reorganises urban space by completely misunderstanding it. Drawing on a striking collection of news reports about burglars who burrowed through walls, roofs and floors to enter buildings, Manaugh argues that the geography of urban spaces can be rethought in radical ways if we put ourselves in the position of the burglar. A burglar could be 'the worm in the apple', as Manaugh puts it, boring a straight line through a city block that's diverted from the normal path of entrances and exits.[10] We might therefore say that, rather than using the more circuitous route of travelling through the door provided for that purpose by the building's design, they take a 'straight shunt' – a direct detour, a shortcut that's also a bypass – right into and through the building. Gately's straight shunt alarm bypass could thus be understood as a direct route, transposed from the physical entry into a building onto the alarm itself.

Manaugh's overarching argument is that burglars creatively misread architecture with 'brilliant stupidity', revealing a 'porous world' where 'everything architects take for granted can literally be undercut, punched through, knocked down, or simply sidestepped'.[11] They reveal that walls and perimeters are not as impermeable as they might appear. To go back to the earlier observation from Brandon LaBelle, both burglars and alarm systems 'perforate' apparently sealed buildings. This pre-existing perforation is emphasised by Gately's burgling modus operandi; since his specialism is defusing alarms, he chooses to operate exclusively in those dwellings whose perimeters have already been bored through by an alarm system. Both burglars and alarms 'straight shunt' alternative paths into contained spaces. If the lesson we took from the representations of the smoke alarm in the previous chapter was that we need urgently to reconceptualise risk to make safety a collective endeavour, rather than just an individual responsibility, representations of the burglar alarm make a similar move in bursting open the closed space covered by the alarm. Gately's 'straight shunt' reveals how the security architecture of the burglar alarm relies upon calling attention from the outside to the inside, rather than sealing spaces completely against intrusion. This makes sense, I think, because alarms are attention extensions: they supplement our attention both in time (as in alarm clocks, which are alert *when* we're not) and in space (as in security alarms, which are alert *where* we're not). Attention naturally extends away from and beyond ourselves, and the alarm represents our desire to cast that vigilance out still farther, with technological additions and expansions.

*   *   *

We can contrast the straight line of the burglar's plotted course with the trajectory of another kind of crime: the heist or caper. If burglary involves busting through the boundary defined by the alarm (by disabling or breaking it with the straight shunt) then the heist more often imagines an elegant and baroque digression around, over, or through the alarm's zone of protection. As a genre in fiction and film, the heist offers a way of thinking about the satisfactions and pleasures of the highly planned and perfectly executed circuitous route. The heist – in theory at least – departs from the repeatable formula of the 'straight shunt' and instead imagines a convoluted and indirect approach to the bypassing of alarms to break and enter.

Let's return to the Night Fox: the thief's dancing progress presents the plot of the heist distilled into a physical movement – it's stylish, it's intricate, and it's constrained by security measures. The scene operates through and around a maze of lasers, set up as motion sensors to protect the art in the museum. I am fascinated by the laser maze trope, which seems in equal parts preposterous and beguiling. It is a set-piece that has its absolute high point in *Entrapment* (1999) – a film which made the laser maze so integral that it featured on the movie posters. In the film we watch Catherine Zeta Jones twisting and pirouetting first through a practice room strung with red yarn and little Christmas bells, then through the real maze to steal an antique from an art museum. *Under Ten Flags* (1960) is, I think, the earliest example of the laser maze. In this film a naval intelligence officer attempting to steal some Nazi shipping codes rolls and tumbles through the maze with his tie tucked into his dress shirt, while wearing the special glasses that make the room's 'invisible net' of infra-red beams visible (to both him and the audience). The acrobatic thieves who caper through the laser mazes avoid the security alarm, searching out the pockets of space where they can evade the attention of guards and watchmen.

As a security measure, the laser maze is pretty unworkable (a curtain or full saturation of sensors would obviously be more effective than a field of lasers, which is full of burglar-sized holes), but these films expect me to suspend my disbelief, just as I ignore my knowledge that real passive infra-red (PIR) motion sensors are not visible to the naked eye. A more realistic silent alarm (notifying only security staff), triggered by an invisible PIR sensor, has none of the drama and thrill of the caper through the laser beams. This is impeccable security theatre, then, in which performance must outstrip practicality.

On film, the laser maze offers a stylised and artificial visual representation of alarm evasion. It's a permeable barrier that's been constructed, really flagrantly, for plot purposes. There's a famous explanation of plot from the narrative theorist Peter Brooks, which suggests that one of the pleasures of plot is deferral, of not arriving at the destination too quickly. Brooks describes plot as 'a kind of arabesque or squiggle towards the end' in which detour, digression and diversion are a central part of the pleasure of moving through a story towards its final conclusion.[12] If a plot were simply to trace a straight line (what Brooks identifies as 'the shortest distance between beginning and end'), then the narrative would collapse altogether.[13] A plot means imposing some measure of artistry onto events by organising them into a satisfactory order, with delays and digressions essential to the deferred pleasure of the conclusion. The heist therefore inscribes the arabesque or squiggle of its satisfying plot onto the physical trajectory of its characters. The other name for the heist genre, the caper, perhaps captures this spatial or physical metaphor more effectively. Skipping, leaping or dancing, rather than walking in a straight line, the caper makes playful progress through space. The laser maze thus creates a physical caper that carries the action forward, with alarms – the little bells on the string, the motion sensors – determining the shape of the plot.

Julian Hanich reads the heist as a genre that holds audiences enthralled by characters' displays of physical competence, technical know-how and sheer audacity; it's a popular genre that, he argues, 'admires skilful virtuosity'.[14] Heists delight because it's wonderful to see characters having a charmed relationship with their physical existence, where tools and technologies work smoothly in their hands, and their bodies move gracefully through the spaces around them. Because of these properties, Harich remarks that it might be instructive to think of the heist as the opposite of slapstick or physical comedy. A caper is what happens when things go right, a pratfall when they go wrong; the bumbling burglar is the foil to the graceful thief. There's also an argument from film scholars that the heist is a self-referential genre that's interested in representing a vision of how films are made. Both a heist and a movie bring together a whole group of people with esoteric but finely honed skills to produce something spectacular. The film scholar Kim Newman argues that the heist film is perhaps closest to the 'putting-on-a-show' musical in its interest in the processes of creation and its potential for self-referentiality.[15] We can therefore understand the heist as a genre that offers a metacommentary on how art gets made and where aesthetic pleasure and satisfaction come from. But it's also a genre which places the alarm's vigilant form of attention at its centre.

*   *   *

The heist reminds us that a squiggly line is more interesting than a straight line, and that skilful planning and execution are worth watching in and of themselves. If the heist film represents a team of highly skilled operators planning and executing a plot, then we might say that what the heist film celebrates is a vision of art not as the product of an individual genius or hero, but as a powerful act of co-creation. This view of the artist is also why it's so significant that many heist films take the art museum or gallery as their setting. In contrast with the individual genius of the great painter, cinema is a form that has to acknowledge collaboration and sociability. One of the ways in which heists offer a metacommentary on art is through the particular subgenre of the art heist.

Many art heist films involve spectacular set-pieces that depend on art museum security systems: recall the Night Fox's journey through the laser maze, the steel security shutters in the 1999 remake of *The Thomas Crowne Affair*, or the playful use of the boomerang to disable the alarm system in *How to Steal a Million* (1966). Setting a heist in an art museum allows for some critique of the security cultures of gallery and museum spaces themselves. Real guidelines on security for galleries and museums emphasise an approach to alarms which involves multiple layers of protection: first, perimeter alarms on windows, doors and vents; next, heat detectors or infra-red sensors to saturate internal spaces with motion detection; and finally vibration sensors to alarm specific works.[16] Moreover, the design of the exhibition space itself – its sightlines, the placement of guards, gallery attendants, and invigilators, monitoring by cameras – constructs a space that's charged with a kind of attention that is not purely aesthetic. Among this thicket of surveilling attention, we're meant to be looking at a Monet.

The art museum's mission is to exhibit art in public view, but that also means placing a highly securitised limit around the kinds of engagement with the artwork that are acceptable. Museums need to protect the exhibits, and this means enforcing a particular route through their spaces and limiting contact to looking and not touching. On the other hand, we could also see ourselves, watching the heist, occupying the position of an overseeing security guard or watchman, observing the action as though on CCTV. Even as we might sympathise with and celebrate the thief's skilful alarm bypassing, the screen makes us inhabit a version of the vigilant attention that the thief is trying to avoid. To me, this recalls the little house made by Edwin Holmes to market his security alarms, through which the customers are invited to imagine themselves as both burglars and watchmen, both breaking in and responding to the alarm.

A tension therefore appears between the security-bounded, aesthetic attention that the museum ostensibly encourages and the alarm-extended, watchful attention that's present in its security systems. The thief can operate outside of both. The art heist imagines what is perhaps my fantasy museum visit, in which I can get as close to the objects as I like and can handle the works without the security restrictions of the institution. If I'm rooting for the art thief, I might also be rooting for an encounter with a work of art that's outside the permitted bounds of the museum's secure organisation of space, and outside the kinds of attention managed by the alarm.

Daryl Lee argues that the heist film is a place where 'the business of film tries to work out its relation to art',[17] and we can conclude that the art heist uses the alarm to identify precisely that tension between business and art, economic value and aesthetic value in the museum. If the artwork is stolen, it goes from public display into private ownership. The Monet in *The Thomas Crown Affair* or the Fabergé egg in *Ocean's Twelve* are protected by alarms and heat detectors and metal shutters because of their economic value, but also to preserve access to them for other visitors. At the root of these contradictions, I think, is the way that the art heist recognises that the alarms in the museum are a form of exclusion that maintains access; they keep people out in order to let people in, create private spaces in order to allow artworks to be available in public view, and acknowledge the tension between economic and aesthetic value. It's this understanding of how security alarms complicate the relationship between inner and outer, private and public that we also observed in the account of the burglar alarm in *Infinite Jest*.

In every example we've encountered in this chapter – from Edwin Holmes's connection of the early burglar alarm to the telephone network, to the burglar's 'straight shunt' and the heist's capering disruption of private ownership and public access – the security alarm makes the relationship between the alarmed zone and its exterior more, not less, complicated. What ought to prevent entry must also expect it. As alarm imaginaries in fiction and film bypass, break, and breach the alarmed perimeter, they model the movement of the security alarm itself which, like every alarm, extends itself through and beyond the boundaries that it purports to maintain.

4 SIREN

Theorists of sound consider the siren a technology that's emblematic of modernity. The philosopher Peter Sloterdijk writes that sirens are 'the bells of the industrial and World War age', suggesting that the bell (and its associations with the peace of rural life, small communities, and the regular tolling of local time) has been replaced by the siren (associated with the rhythms of urban factory work and the horror of total war).[1] For Sloterdijk, the 'sonosphere' of the siren is one of 'forced alarm', of 'terror' and 'panic'.[2] This 'most brutal of mass signals' can only convey a negative message, a 'consensus that everything is hopeless and dangerous'.[3] His reasoning is that, unlike a bell which could peal in celebration as well as ring out a warning, a siren is purely a sound of alarm. Sloterdijk's sonosphere is a bubble of sound, inside which a group of people share a common feeling: jubilation when the church bells ring; panic when the siren sounds.

The sound theorist R. Murray Schafer shares with Sloterdijk a historical conceptualisation of the siren that contrasts it with the bell, identifying both as 'community signals' with different emotional associations: 'the shape of the bell suggests community', he argues, while the siren 'broadcasts distress'.[4] Both the bell and the siren radiate sound uniformly in all directions, creating a circle of reach on all sides. (In this sense they're distinct from horns and whistles, which are directional noisemakers.) Schafer's analysis of the siren focuses on the alarm mounted on an emergency vehicle which, he argues, 'is designed to scatter people in its path'.[5] In these accounts, the siren reaches masses of people and evokes the same feeling in them. It moves them both emotionally and physically, making them scatter and disperse.

Sirens interest me because they represent the loudest possible alarms, reaching physical limits of sound. The loudest ever siren is listed in the Guinness Book of World Records as the 1952 Chrysler Air-Raid Siren, which was the model that would be rolled out for Civil Air Defence in the Cold-War United States. The volume was measured at 138 decibels 100 feet from the siren, and the sound could be heard from four miles away.[6] Earlier materials from Chrysler had boasted that their air-raid sirens were loud enough to disperse fog, and the Guinness record claims that the 1952 model was so loud that it could ignite a piece of paper held in front of the siren horns.[7] In photographs, the Chrysler air-raid siren looks more like a little red locomotive from a miniature railway and descriptions of its action in its operating and maintenance manual make it sound like a threshing machine or a woodchipper, sucking air in and separating it out into chunks of sound with the 'chopper'.

An ambition to construct the loudest possible alarm connects the siren with the foghorn, another device for creating enormous sounds. In Jennifer Lucy Allan's beautiful book about her long fascination with foghorns, she describes nineteenth-century attempts to create a sound that could travel miles across the sea. She notes that the foghorn is 'technologically, but also functionally' different from the siren.[8] Where, for Allan, the siren is a 'call to action, to move' – like Murray Schafer she thinks of the siren's alarm as a warning to scatter – the foghorn is an aid to navigation; it's a notification given to the ear when the eye is impeded by fog.[9] Nevertheless, the foghorn and the siren both developed from experiments with powered pneumatic horns that could produce louder and louder mechanised sounds. These devices in turn came from the musical instruments of the reeded horn and the pipe organ. The largest foghorns that Allen describes in her book do look musical, with their 'big trumpet mouths'.[10] But until the development of the electronic siren in the 1960s, both the siren's variating wail and the foghorn diaphone's long, low honk were produced by compressing air and letting it loose. They were machines for contorting air into noise.

One version of the siren, then, is a device which uses the loudest possible sound to cover the largest possible area, reaching every person within it. In its loudest forms, it takes air in and fires it out, reaching a universal audience, saturating the area in alarm. Sirens purport to carry a universal warning which will find, as Sloterdijk puts it, 'all ears that can be reached'.[11] In this analysis, the only way to avoid the siren's message is, like Odysseus in his encounter with the mythological Sirens, to stop up one's ears.

If the siren represents modernity, and a departure from the tradition represented by the church bell, then we might find that the age of the siren, too, has passed. Michael Bull adds an important addendum to this historical account of the siren by identifying more recent tendencies towards the 'privatization of warning' through the use of individual alerts sent to mobile phones.[12] As I was writing this book in the summer of 2021, the UK government began to trial its first emergency alert system, which was adopted in response to the Covid-19 pandemic. The system involves sending messages to mobile phones within reach of particular masts within the phone network. In late June and early July, some phones were sent test alerts, signalled by a 'loud siren-like sound' and accompanied by a text message explaining that this was a test of the service.[13] Each mobile device therefore becomes something 'siren-like'; an object that's enough like a siren to register all of the ways that it's not a siren at all. The mobile phone emergency alert is individualised, small-scale, and only reaches those with a phone that's turned on and in range. It doesn't have the same ambitions for universality as the siren, and its sound, supplemented by text, can be much more specific and unambiguous.

Analysis of the siren has made two consistent claims about it: that it sends a universal message through air's neutral medium, and that it scatters people in its path.

Nevertheless some of the most powerful cultural objects that explore sirens are interested in the ways that these alarms do not convey a universal message to every ear. Instead, they find that the siren – and the police siren in particular – is an alarm that carries a message of reassurance and protection for some while sounding out an urgent threat for others. Moreover, even as it disperses those it protects, the police siren arrests those it criminalises. As we'll see, the work of Black writers, artists and musicians has explored the reach and limitations of the police siren, and re-sounded the siren as a discriminating object rather than a universal call to 'all ears'.

*   *   *

Sirens appear and reappear in Black art as signifiers of the ways that racial categorisation is constructed and maintained through policing. KRS-One's 1993 track, 'Sound of da Police', for example, gives voice to the sound of the police siren (woop woop) in the lyric, making it re-sound with new meaning.[14] 'Sound of da Police' emphasises that the siren does not scatter everyone with a universal call. That universality may be true of an ambulance or fire engine siren, but the police siren, while it scatters some people, for others has the opposite physical effect: arresting them, stopping them and fixing them into place. Not everyone has the same experience of this supposedly universal siren, with some hearing it as call to move along and others hearing an order to stop. This differentiation is, as we'll see, shaped by racism and police violence.

The onomatopoeic woop woop in KRS-One's lyric is specifically the short yelp of siren noise used to indicate that a driver should pull over, rather than the longer siren wail more commonly used for clearing traffic. With the adoption of electronic sirens for emergency vehicles from the 1960s onwards, the new capacity for variance in tones meant that police sirens could take on more nuanced meanings. In 'Sound of da Police', drawing people physically to a halt is a crucial instrument for racialised monitoring and state violence. As the track establishes through its punning affinity between the words 'overseer' and 'officer', the siren is the sound of racist harassment which draws a straight line from the violent surveillance of chattel slavery to the stop-and-frisk tactics of modern policing. Both function by stopping people from moving: the lyrics describe how the overseer would make enslaved Africans stop what they were doing, while the modern police officer pulls over Black drivers to subject them to a surveilling attention. Like all alarms, the woop woop of the siren expects a response: it calls its Black subjects to a standstill and places them into a position that's weighted with a history of racist violence. By revoicing the siren in the lyric, KRS-One sounds out that history and makes the siren's discriminating instruction – 'you: scatter; you: stop' – impossible to overlook.

There's a second important trope associated with the police siren which has a longer history in Black art. This establishes the siren, the loudest of alarms, as a device that also silences. Richard Wright's classic novel *Native Son* (1940) escalates the sound of alarms throughout the narrative, reaching a climax of overwhelming police sirens when its protagonist, Bigger Thomas, is about to be apprehended by police. In the text, the descriptions of the sirens and their effect pair the sound and a specific bodily affect:

> He listened; there were throbs of motors; shouts rose from the streets; there were screams of women and curses of men. He heard footsteps on the stairs. The siren died and began again, on a high, shrill note this time. It made him want to clutch at his throat; as long as it sounded it seemed he could not breathe.[15]

Reading this passage, Jennifer Lynn Stoever finds that the 'policing noise' is not just background sound; it functions, she argues, 'as a sonic lynch mob' surrounding Wright's protagonist.[16] For Stoever, and I think for most readers of Wright's work today, it's difficult not to also hear in Bigger's struggle to breathe the final words of Eric Garner and, more recently, of George Floyd – men who died by choking and suffocation at the hands of police officers.[17] Stoever finds that Wright associates the police siren with choking and with silencing, drawing on the specifically racial history of lynching that is evoked by the siren's suffocating sound and its pairing with the shouts of a crowd.

In an essay, 'How "Bigger" was Born', which is now customarily appended to the novel as a foreword, Wright describes his writing of the novel and some of the challenges it presented. In the essay, he recounts his difficulties with the novel's opening scene for which, he says, he had identified a particular emotion that he wanted to convey, but struggled to find a 'concrete event' that would serve to carry 'in varied form, the note that was to be resounded throughout its length'.[18] The title Wright gave to the first part of the novel, 'Fear', gives some indication of its dominant emotion, and he did finally settle on an opening event: a famous scene in which Bigger kills a rat with a skillet in the overcrowded room he shares with his mother and siblings, and during which his younger sister faints in terror. But before this event, the novel begins on a much more literally resonant 'note' with sound of an alarm clock, written as an onomatopoeic 'BRRRRRR RIIIIIIIIIIIIIIIIIIINNG!'. This opening alarm, if understood as Wright's 'concrete event' that resounds through the novel, is the start of an escalation of alarms that culminates in the police sirens. In the later scene of Bigger's pursuit, the novel describes his position, hearing voices 'under the terror-song of the siren'.[19] The opening sound of the alarm therefore creates the resonating sonic disturbance that structures the rest of the novel and makes terror and fear its dominant emotions.

The imagery of being 'under' the siren – and specifically under its affective sweep of terror – associates the siren's regime of feelings with the historical violence of slavery that also incorporated strict processes of surveillance and watchfulness. As in KRS-One's association between the officer and the overseer, the imagery of being 'under' the siren also recalls the overseeing vigilance of a surveilling attention that can be accompanied by lethal violence. Simone Browne's book, *Dark Matters*, offers a history of the processes of surveillance practised by white overseers, but also of the practices of 'dark sousveillance' that meant those processes could be 'appropriated, co-opted, repurposed, and challenged' to ensure the survival of enslaved people.[20] Sousveillance, observing from below, could include the handbills distributed by abolitionists raising the alarm for people liable for prosecution under fugitive slave laws to avoid 'watchmen and police officers' and 'keep a sharp lookout for kidnappers and have top eye open'.[21] But it could also include the courage of Darnella Frazier, the 17-year-old girl who used her phone to record the ten minutes and nine seconds of footage which documented the murder of George Floyd. Works such as 'Sound of da Police' and *Native Son* are also acts of dark sousveillance which record and rework the terror-song of the overseeing siren from underneath its surveilling sonosphere.

We can understand these representations of the police siren to be making a distinction between those who are protected by the law (and who scatter) and those who are halted, suffocatingly, in place to become subject to policing. They convey how the siren – the loudest alarm – stops the voice, prevents breathing, and connects modern state violence with a longer history of racist violence. But they also offer a way of disrupting or reimagining the siren from under its terror-song. Re-voicing the woop woop or the wail reveals new resonances to the siren, and turns an acute aesthetic and political attention back on its overseeing surveillance.

* * *

Claudia Rankine's book-length poem, *Citizen: An American Lyric* (2014), also draws on patterns of imagery that associate sirens with a history of halting or apprehending Black Americans, and of silencing and suffocation. Rankine's book mixes together essays, images, scripts for short videos, and prose poetry in a self-reflexive lyric voice that addresses itself as 'you', and offers observation, warning, and contemplation. One of the threads that runs through the poem is an attempt to convey the patterns of police violence, culminating in a list three-quarters of the way through the book of Black citizens who have died because of contact with the police. In each new edition names are added and Rankine's memorial list continues on down the page with spaces waiting to be filled as the text fades out and disappears.[22] My 2015 edition stops at the name of Sandra Bland, who was found hanged in a police cell in June of 2015 after she was arrested following a traffic stop three days earlier.

Citizen uses the siren to anchor together different incidents of apparently minor racist slights with the constant background awareness of police violence. The book starts with a catalogue of microaggressions and cruelties – the friend who, distracted, calls the speaker by the name of her housekeeper; the little white girl on the plane who is reluctant to sit next to her – and running through these early scenes is the sound of the police siren. In one of these vignettes, the speaker receives a phone call from a neighbour reporting a 'menacing black guy' outside her house.[23] Quickly, it becomes clear that the guy is a friend, dropping off the speaker's child after school while she goes to the cinema. Hearing 'sirens through the speakerphone', she realises that her neighbour has called the police. The film she had been to see was *The House We Live In*, the documentary about the prison system, the war on drugs, and the criminalisation of Black men. This is the first of many examples in the poem of the siren signalling contact with the police which exposes the speaker, her partner, and her friends to danger. The siren becomes the marker of incoming threat; an alarm indicating the moment when racist slights could easily escalate into state violence.

The siren first appears in the book as a fantasy, as the speaker imagines a pretext for bringing a racist conversation about hiring practices to a halt, wishing 'the light would turn red or a police siren would go off' and let her slam on the brakes.[24] Rankine takes this association between the siren and the stop and sets it in dialogue with a second command from a culture which demands that Black citizens should let go of the past, stop dwelling on history, brush off discrimination and keep moving. Towards the end of the poem, the speaker coaxes herself and her partner back into driving after being stopped by the police:

> Come on, get back into the car. . . .

> Yes, and this is how you're a citizen: Come on. Let it go. Move on.[25]

Even as Trayvon Martin's name 'sounds from the car radio', the speaker is submitting to these two conflicting demands: to stop at the sound of the siren and to 'move on' and 'let it go'.[26] Where, at the start of the poem, the speaker had heard the sirens over the speakerphone, now she's hearing the consequences of police violence over the radio while she hears the sirens in person. As in Wright's *Native Son*, the emotional atmosphere of this text operates 'under the terror-song of the siren'.

The most striking appearance of the siren in *Citizen* appears in the script for one of a series of 'Situation' videos.[27] In this piece, titled 'Stop-and-Frisk', the siren is a part of the standard vocabulary of a traffic stop:

> Each time it begins in the same way, it doesn't begin the same way, each time it begins it's the same. Flashes, a siren, the stretched-out roar –

> . . .

> And you are not the guy and still you fit the description because there is only one guy who is always the guy fitting the description.[28]

In 'Stop-and-Frisk' these portions of text recur, with some small variations. This is the routine repetition of the procedures of the traffic stop, but it is also the repetition of trauma.[29] Reading these repetitions through the logic of the alarm is also instructive. We've established that alarms function as part of repeated processes of preparedness, of routines and drills. The conventions of the traffic stop, beginning with the alarm of the siren are, by necessity, drilled into Black children. The routine repetition of the traffic stop, then, operates by the rehearsed logic of the alarm. The encounter is choreographed ('Get on the ground', 'Put your hands in the air') and scripted with standard lines. However, the speaker isn't able to play the part:

> Yes officer rolled around on my tongue, which grew out of a bell that could never ring because its emergency was a tolling I was meant to swallow.[30]

There's another alarm here – a bell, not a siren – that can only sound out a silenced, internal emergency. In this passage there's good evidence of what Shermaine M. Jones describes as the poem's 'affective asphyxia' in its choking down of pain and outrage and its emphasis on the bodily consequences of everyday racist encounters.[31] The suffocating, silencing sound of the siren that we identified in Wright's novel scales down here to this emergency bell, its tolling swallowed and swallowed again.

Citizen as a whole works by repetition, but a repetition that's not the same as a preparedness drill. Instead, experiences of routine discrimination and prejudice accrete into life-halting and suffocating racial trauma that make it necessary to reject the comfortable speaking 'I' of the conventional lyric voice. As Sandeep Parmar writes in praise of the poem's handling of voice, *Citizen* does not sink back into a 'universal', 'self-assured' poetic I – or voice an othering and 'ornamental' difference.[32] Without repeating the falsely universalising sound of the siren, the poem's voice is nevertheless urgent and alarming.

These works examine what it means to be the subject of the alarm's prosthetic attention. In its overseeing surveillance, the police siren indexes those aspects of the alarm that construct a supplementary, extended attention that goes intrusively farther, augmented by a prosthetising technology. But in the texts considered in this chapter, we've witnessed Black artists directing attention back to the siren as a tool of overseeing vigilance and racist violence by re-sounding it with the very voices it silences. These works short-circuit the siren's oppressive and over-reaching attention. Ultimately, the lesson we should take from the siren is that alarms are not universal or neutral technological devices; they're social and cultural objects whose meanings are derived from the world that surrounds them, and whose meanings can therefore be remade through creative reappropriation.

5 FAILURE, FALSE, FATIGUE

Alarm failures can help us understand more about what exactly alarms promise and what happens when they do not give us what we'd hoped for. False alarms happen when alarms are too sensitive or not sensitive enough, while alarm fatigue occurs when those operating or responding to an alarm have become desensitised (often because the alarm is, in turn, too sensitive). The argument I've been making all through this book is that alarms are objects that regulate and administer our own feelings by feeling *for* us, by taking over some of our capacity for alertness on our behalf. But there are times when alarms seem to feel too much, or demand too much of us. False alarms and alarm fatigue are products of this miscalibration between the sensitivity of the alarm and the sensitivity of its respondents.

Imagine a catalogue of positive alarm archetypes: sentinels and prophets, whistle-blowers and watchdogs. These are heroic figures who identify threats, issue early warnings, and raise the alarm to prompt a community to protect itself. Alongside these examples, however, are a catalogue of archetypal figures who represent a failure in the social processes of alarm: Chicken Little, the boy who cried wolf, the legal exemplar of the person shouting fire in a crowded theatre. These are the doomsayers and fearmongers, the alarmists who endanger others with warnings that exaggerate a threat. We also ought to include a third category: those who raise the alarm and aren't believed, as in the example of Cassandra considered in this book's introduction. Each of these figures serve an archetypal function in identifying the parameters of alarm as a social, rather than a technological process.

Turning to stories that serve as cautionary tales or instructive fables about how to raise or respond to an alarm, this chapter identifies how new kinds of stories can help to recalibrate social processes of alarm. Chapter Three considered some of the reasons why alarms might be ignored and explored how practices of drills and rehearsals prepare us to react to alarms. When the present's overlapping crises of climate collapse, mass extinction, global inequality and social injustice are regularly dismissed by powerful people as false alarms or hypersensitive over-reactions, conceptualising how alarms work can help us to make decisions about how to respond to warnings in a proportionate way. At the same time, as social media filter bubbles radicalise people into conspiratorial thinking, we have a responsibility to envisage new ways to respond to people who are sincerely very alarmed about things which are false. Art can help to do that work of investigation and imagination.

In the previous chapters we've seen that films, music, literature, and visual art can uncover how alarm technology can reflect upon existing social relations. When KRS-One re-voiced the woop woop of the police siren or Francis Picabia printed the alarm clock, they were exposing the workings of these alarms, but also the workings of their own societies – from the racial violence of the police stop to the dulling everyday routines of capitalist modernity. Art about alarms can reveal and reconceive them, as well as activate a sort of aesthetic counter-attention which harnesses the alarm's attention-grabbing properties and turns them back on the object itself. In order to draw out some of the implications of this book's focus on art and culture, this chapter turns away from alarms as objects for a short while and instead considers practices of storytelling which create and critique alarm archetypes.

* * *

We know already that an alarm can only work properly if it's expected – if those who hear it have been trained to know how to respond. Alarm archetypes such as the boy who cried wolf or Chicken Little could be explained by their preparatory function, priming responses ready for the eventualities that might arise from real-world alarms. However, the characters I want to think about here offer not just training in alarm response, but also a kind of alarm consciousness which requires their audience to think critically about alarms as a social process. 'The Boy who Cried Wolf', for example, is a cautionary tale that warns children not to raise a false alarm. The boy experiences the consequences of alarm fatigue in his adult supervisors, giving the audience a lesson in those consequences too. 'Chicken Little' is a warning against believing messages of alarm – the story's panicked warning that the sky is falling – that might be overblown or misconstrued. In this example, child listeners are encouraged to put themselves in the position of Chicken Little's friends, and to advance reasonable scepticism towards other people raising the alarm. Your friends might not be lying, the story explains, but they could be mistaken or misled and, even as they're warning you of huge and dramatic threats, they might be putting you at risk of other more ordinary dangers (like being eaten by the local fox).

In an age of fake news, conspiracy theories, moral panic, and an attention economy that rewards viral content that evokes strong emotional reactions, these fables against alarmism might seem enduringly relevant. But at the same time, these archetypes of alarm are also routinely weaponised to dismiss anyone who might want to warn against threats that are too big or slow or structural or private or unevenly distributed for other people to acknowledge. The woman who 'cries' rape, the 'alarmist' climate scientist or epidemiologist, the 'woke' BLM activist: all of these people are routinely dismissed as false alarms whose sensitivity settings have been calibrated incorrectly. At best they're feeling too much, or mistaken about their evidence; at worst they're malicious liars deliberately raising a false alarm to make other people take actions that aren't in their best interest. The appropriation of the AAVE term 'woke' and its transformation into a pejorative is instructive, I think, as it turns not just the raising of alarm but even the alertness to or awareness of injustice into something worthy of condemnation – even *noticing* racism is objectionable. 'Anti-woke' discourse is an attempt to dull the kind of vigilant attention that's the alarm's hallmark, and to make it less acute, less sensitive.

The false alarm that we encountered at the beginning of this book in the story of Cassandra also raises another question about sensitivity and alarm. In Aeschylus's play, as well as raising an early warning about Agamemnon's death, Cassandra also raises another kind of alarm by telling her story of what sounds like sexual assault committed by the god Apollo, after which he cursed her. Cassandra is therefore condemned to repeat the experience of reporting a terrible wrong, but being dismissed as a liar and a fraud; Apollo turns everything in Cassandra's life into a false alarm. Cassandra's curse of prophecy also, to me, reads as a brutal metaphor for psychological trauma, which sends up hypervigilant alarms everywhere, makes you scared when you should feel safe, and numb when fear might protect you. Trauma throws our own systems of alarm into disarray and results in calculations of risk that seem unreasonable—paranoid, hysterical, callous— to people who don't share our history.

Real false alarms do happen, of course, but the concept of the false alarm plays an outsize role in the cultural imagination. The story of the boy who cried wolf, the disproportionate fascination with the idea of women making false allegations of sexual assault – how can we explain why falseness so readily attaches itself to the alarm? One of the provocations that Sara Ahmed makes in her book *The Cultural Politics of Emotion* is that it's 'the objects of emotion that circulate, rather than emotion as such'.[1] This is the crucial move that allows her to think about how feelings accrue around particular objects as the objects move in circulation. She analyses how the term 'asylum seeker' becomes attached to other words: swamped, overwhelmed, bogus (a particularly significant term in the context of the false alarm). Ahmed suggests that 'signs become sticky through repetition' – they stick to each other as they are used together.[2] The word 'false' is a sticky sign that adheres to the word 'alarm' as it circulates around as an object of emotion. The stickiness of these words also exposes something about what we want systems of alarm to do and be. We want alarms to be a switch that can rationalise alertness, arousal and affect, dispensing them in the correct quantities at the correct moments – to ration them out because they're limited resources. The alarm should cleanly allocate feelings. But the false alarm shows the sticking points of that imaginary mechanism and the failure of alarms to organise feelings cleanly and rationally. Objects of emotion circulate stickily through this system, getting other sticky signs stuck to them, and nothing comes out clean.

Another part of the answer to why false alarms are so disturbing is because of the obligations for action that rest on the alarm. We began this book with a definition of an alarm as a signal which requires a response. That response might be arming yourself to defend the town or rushing to get the wet clothes out of the washing machine. Alarm as a social function therefore becomes a kind of paradigm for the way that words have consequences. The ultimate example of an imagined alarm used in this way is probably the legal exemplar of 'shouting fire in a crowded theatre', representing the limits that must be placed on free speech to keep people from harm. This hypothetical boundary at which freedom of speech becomes materially damaging relies on the idea that an alarm, by definition, demands a response, and that because words have consequences it's reasonable to limit certain kinds of speech. Raising the alarm means turning words into actions, and understanding that words have consequences in the world.

Researchers in alarm psychology suggest that there are two elements necessary for an effective alarm response: compliance and reliance.[3] Both elements are unsettled by false alarms. Compliance is the rate at which people respond when they are faced with an alarm. A high rate of false alarms means that respondents, like the adult shepherds in 'The Boy Who Cried Wolf', will simply not comply with the alarm's instruction. Reliance is the confidence that people can have in the alarm when it is not sounding. Reliance will be high if users trust that an inert alarm is not activating because there's no reason for it. We could imagine another alarm fable, perhaps, in which a boy watching the sheep falls asleep and doesn't alert the other shepherds when there is a wolf attack. The shepherds would find him an unreliable alarm, whose silence couldn't be trusted to mean that all was well. In this situation of low reliance, the alarm would also cease to function as an extension of their own vigilance. Compliance and reliance are interesting to think about because in situations outside technological alarms – i.e. when people themselves raise the alarm – compliance and reliance are also, inevitably, inflected with power and social status.

At the beginning of this book we read the story of Cassandra as a tragedy in which the plot's irony arises from the fact that Agamemnon, the powerful tragic protagonist, was warned about what would happen and ignored those warnings. The audience is invited to witness the consequences of Agamemnon and the elders of Argos dismissing Cassandra's warnings as false. In watching the play, we experience the horrifying feeling of Cassandra's helplessness and vulnerability. The stories of the boy who cried wolf and Chicken Little embed a similar power imbalance in their characterisation, with both stories imagining child characters with only limited authority. They are stand-ins for their child audiences, learning to bear the responsibility of processes of alarm. But, reading them as an adult, it's difficult not to interpret these stories against the grain, and to feel sympathy for an attention-seeking boy or a little chicken overwhelmed by the apocalyptic anxiety that the sky might truly be falling.

These fables of alarm indicate that trust and authority are central to the conception of the successful alarm. As we've established using the example of the domestic smoke detector, the ideal alarm is one that never needs to be used. By simply existing in stoic and silent vigilance, the alarm fulfils its function with a promise of a warning if anything goes awry. In contrast, the malfunctioning alarm is childish, hysterical, hypersensitive, or untrustworthy.

In guides for engineers about managing alarms, one of the striking things is how full of character the bad alarms are: they 'chatter' if they're oversensitive and need acknowledgement; if they tell you things you don't really want to hear then they're a 'nuisance'; they're 'bad actors' if they present you with the wrong information. Here's one cute example of a piece of imagined dialogue between a nuisance alarm and the automated control system meant to manage it:

NUISANCE ALARM TO CONTROL SYSTEM: Hey! I am in alarm!
CONTROL SYSTEM: Yeah, I get that a lot from you. I am going to wait – let's see – ten seconds before I tell the operator.
NUISANCE ALARM: *8 seconds later* Hey! I am not in alarm anymore!

CONTROL SYSTEM: Yeah, I figured that would happen.
Good thing I didn't bother the operator with your
prior message.[4]

With its limited capacity for communication, the alarm takes
on the role of a bratty, attention-seeking little kid, while the
control system is cast as a sort of algorithmic nanny, ready
to put the alarm in time-out until it can stop playing up.
There's utility in thinking of alarms in personified ways – as
shrieking or nagging or grumbling – and these attributions
certainly tell us something about how we feel about alarms
and our responses to processes of alarm more generally. But
this personification also has its limits. Alarms might be like
people (because we made them) but people aren't always
like alarms. There are forms of alarm response that can only
operate outside the mechanical or algorithmic logic of the
alarm, and which stories can help us to explore.

*   *   *

Lydia Davis's beautiful and moving short story, 'Fear', reads
to me like a corrective to 'Chicken Little' or 'The Boy Who
Cried Wolf'.[5] It imagines what would happen if we were able
to give people exactly the care and attention that their alarm
demanded, and to go beyond the mechanical allocation of
appropriate feeling into something messier and more human.
    In this very short story of only four sentences, the narrator
describes how a local woman regularly runs out into the

street, clearly distressed, and shouting out an alarm for an emergency. The other people in the street know that there is no danger – the narrator tells us that they all know that she is inventing the emergency – but people in her community rush to give her exactly the comfort and attention that she needs to calm her. The story's long final sentence sees the narrator describing, in a sympathetic reflection, how the woman's neighbours console her because they have felt the urge to do just the same, and when they had those experiences themselves they needed to seek help from the people around them to console them. Seemingly without friends and family to comfort her, the woman in this community needs the support of her neighbours to quell her fears.

Like Chicken Little, the woman believes her fears to be entirely real and feels compelled to alert others. Like the boy who cried wolf, she risks the alarm fatigue of the people around her with her daily false alarms. We know the archetypes already, so when we read the first sentences of the story we might expect that her neighbours will be united against the alarmed woman, or pulled into her anxious alarmism. But the final long sentence shows the community acknowledging her fear, and recognising the feeling in their own past, even while they know there's no emergency. Rather than being punished because of her unfounded or excessive demands on her community's care and attention, the alarmed woman is instead met with exactly the response that meets her needs. Unlike the fables of alarm that teach children to condemn the attention-seeking boy or the hysterical Chicken Little, 'Fear' shows the narrator modelling exactly where the community's empathy comes from: from their own experiences of distress and the aid they've been given by others in the past.

The story therefore invites the reader to take up two positions: first the calm neighbours and then, following the path of their caring attention to her, the alarmed woman. If we recognise that, like the neighbours, we too have felt afraid, and have needed to seek comfort and make unreasonable demands on other people to calm ourselves, then we also have to empathise with the alarmed woman. The little narrative walks us through an exercise in care and empathy.

Alarms always promise both sides of an affective pairing: both security and agitation, both peace and panic. And crucially, the assurance offered by the alarm is that each of those states of feeling will occur only when, like the alarm itself, they are designed to occur. The alarm is mechanism that promises the regulation of incompatible affects, neatly allocating the right feeling at the right moment. The peril presented by a false alarm is a disruption of the alarm's promise of regulated feeling, so that emotions are dealt out inappropriately and end up in the wrong time and place. The opposed feelings of Davis's story (the woman's alarm and the community's calm) don't conform to the alarm's idealised regulation of attention and affect. Instead, the characters are moved to respond in an emotional way to the woman's distress, offering unlimited care to a person who seems like they need it. Rather than rationing out their attention and ignoring the alarmed woman's cries, the neighbours meet her demands for their caring attention.

'Fear' is a story that cuts off the supply of anxiety that fuels a preoccupation with the false alarm by suggesting that human feelings and caring attention don't have to be rationed; they can be excessive and extravagant and unreasonable and respond to excessive and extravagant and unreasonable demands. By rewriting these archetypal alarm fables, Davis suggests that attention and feeling outside of the protocols of alarm logic need consideration too, and might help us to escape from at least some of our anxieties about the false alarm.

In contrast with vigilance or alertness, care is a form of attention that is not and cannot be ordered by alarm technology or by the rationalising logic of alarm processes. The various alarm artworks that we've encountered so far bring home some of the limitations of alarm logic which makes the forms of attention supplemented by the alarm (such as vigilance and alertness) into the most important forms of attention. In the example of the art-museum heist movie, we saw how aesthetic attention can operate within and against the monitoring attention of the security alarm. In the work of Black artists and writers, discussed in Chapter Four, we encountered a sousveilling counter-attention emerging from under the siren's overseeing 'terror-song'. 'Fear' reworks alarm archetypes to offer another, different kind of alarm response, which meets human alarm with care and recognises those places where human attention is richer and more varied than the narrow vigilance extended, technologically, by the alarm.

6 FUTURE

This book has argued that alarms are objects that extend our attention and regulate alertness in ways that feel safe, orderly, and efficient. I've suggested that we invest alarms with a collection of feelings about our capacity for attention and our worries that we might not be able to manage our responses to a demanding world. But alarms make demands of us too. They hold up an ever-vigilant ideal for attention that we'll always find ourselves failing to reach. It's these feelings of falling short that I'll return to in this final chapter, as I consider a type of alarm that promises to regulate individuals, their habits, behaviour, and even their physiological and cognitive processes.

In contrast with the largest and loudest alarms – the sirens of Chapter Four – the tendency of future alarms is not towards increasing volume or universal reach. Instead, the future alarm is personal. It's personal in its individualised messaging (as in the example of the text-message emergency alert system also mentioned in Chapter Four) but it's also personal in its monitoring. From earbuds which monitor heartrate and deliver a voice alert if it gets too high, to a wearable thermometer that monitors basal body temperature to notify you of ovulation, devices for telling you about your own body have proliferated until it seems there are few biological processes that don't yield real-time data that can be used to optimise performance or wellbeing. Devices such as smartwatches and wearable activity trackers have become ubiquitous in the last decade and, while they borrow some of their everyday familiarity from the wristwatch, they notify users of more than just the time. Such devices often integrate other alarm functions, such as email notifications or clock alarms, with continuous monitoring of heartrate and motion. A reminder about your step count can appear in the same place as a calendar alert for your next appointment.

Many of these consumer monitoring and tracking devices use sensors to gather data on measures that indicate alertness, activity and arousal. For example, wrist-worn accelerometers can be used to gather data on physical movement as a proxy for identifying periods of lighter sleep, while heart-rate monitors (whether on a chest band or a wrist sensor) are used to identify physical arousal and relaxation. In one device, the Ōura ring, these measures are gathered together under the term 'readiness' – a word that resounds with the alarm's own alert vigilance. Arousal and activity – the most common states tracked by these wearable sensors – are exactly the properties that, historically, have been marshalled and organised by the alarm.[1] These devices therefore take the alarm's processes for extending or delegating attention and apply them to attention itself. Where people once lodged their anxieties about the limits of their own vigilance in torch beacons, buckets full of pebbles, little grains of radioactive material, laser mazes and very big horns, today's attempt to manage a perception of deficient attention is best represented by a watch that vibrates if you've been sitting down too much.

The development of personal, wearable devices turns alarms inward by extending our attention to the processes of the body and its internal states, but also by giving patients more responsibility for monitoring themselves, rather than leaving the interpretation to physicians alone. In a medical context, the consequences are positive, with more autonomy for patients, especially those with chronic illnesses such as diabetes. But the proliferation of consumer devices in the more nebulous field of wellness monitoring has the potential to turn their users into anxious and hypervigilant self-managers, with their attention focused closely on their own bodily processes and performance.

Consumer devices monitor the body, but they also recruit their wearer into practices of alert self-monitoring. The purpose of the Lumo Lift, for example, (a device which sends an alarm message to prompt its user to stop slouching) is not to relieve me of the burden of paying attention to my posture; it's to *make* me pay attention to my posture. In contrast with the alarm, whose promise was to take over the responsibility of vigilance from me, to stay alert on my behalf and allow me to slip into unconsciousness, distraction or diversion, the watchword of these body monitoring devices is mindfulness – the fostering of a constant, low-level, self-improving vigilance. The Lumo Lift isn't really training my back; it's training my awareness of my back. It is teaching me a form of self-monitoring attention as the route to self-improvement.

*   *   *

The wearable activity monitor and its potential for both surveillance and self-surveillance has been the subject of a cluster of recent dystopian fictions: Dave Eggers's 2013 novel, *The Circle* (in which workers at the eponymous tech-giant wear sleek silver bracelets synced to an ingested sensor) and its 2021 sequel, *The Every*; Robert Hart's *The Warehouse*, from 2019 (where 'Cloudbands' monitor work-rate and location and sound an alarm if they're removed for too long); and Joanna Kavenna's *Zed* from the same year (where 'BeetleBands' log everything from tooth-brushing to bad language). It's not hard to see why the wrist-worn body monitor has become such a mainstay of this corporate-tech-dystopian micro-genre, especially after two patents granted to Amazon for wrist-monitors for warehouse staff in 2018 took the monitoring of mega-corporations' workers out of the realm of speculation and into reality.[2] All these novels see the wearable alarm as a representation of how the demands of corporate capitalism are enacted onto the body through a requirement that users should develop their own monitoring forms of mindful alertness.

Dystopia is another genre – like the tragedy, the heist, and the cautionary fable – that is full of the social processes of warnings and alarms. The genre often has a monitory function, identifying tendencies within or trajectories out of the present and asking its audience to make a comparison between its projected future and the world around them. Imagining the future alarm inevitably doubles up on the same early-warning mechanisms.

Of these tech-corporation dystopias, Eggers's depiction of the Circle in particular understands future alarms to be producing a new kind of self-consciousness in the people who use them. As focalised through Mae Holland in *The Circle* and, later, Delany Wells in *The Every*, the workers on Eggers' dystopian tech campus are encouraged to participate in acts of extravagant self-monitoring and auto-administration as the corporation advances an increasingly sinister strategy of 'completion' or 'closing the circle' in which every part of life will be rendered completely transparent, measurable, and productive.[3] Mae is the innocent new starter of the first novel and will become The Circle's CEO in the second. When she is first inducted into the company she is constantly aware of the privilege of her job and of the demand that she should be expressing enthusiasm and passion for her work. Her wrist monitors (and the headset and body cam that follow later) therefore become an extension of the way she monitors her tone in emails, attends social events on campus when she'd rather be alone in her kayak, and keeps smiling in the office even as a 'wave of despair' gathers inside her.[4] She understands the expectation that she surveil herself and submit to the corporation's demands for her to regulate her feelings to fall in line with the firm's goals. And she understands that every part of her life – her working hours, her socialising, her consumer preferences – will be mined for profit through the Circle's insistence that she share them all the time.

The experiences of the Circlers accord closely with the account of the attention economy that appears in Jenny Odell's 2019 book *How to Do Nothing*. Odell argues that the way in which contemporary capitalism demands that we regulate our attention, to direct it always towards productivity and progress, makes doing nothing a subversive act.[5] We've encountered a similar argument before, in Chapter One, with Jonathan Crary's account of the colonisation of sleep by 24/7 capitalism. In both cases, those pockets of non-productive time in which attention and alertness are not directed in the interests of capital – for Odell, birdwatching; for Crary, the moments before sleep or upon awakening – become one of the last places where alternative ways of living might be found.

In Eggers's *The Circle*, we see these little pockets being systematically eroded. Mae begins to check her emails in bed, then moves on to allowing a live feed of a camera to broadcast her sleeping, in order to gather data to help improve her sleep. At the start of the novel, she loves to go to the California coast and watch the birds and seals, and situate herself above the great opaque depths of the ocean. She's practising exactly the forms of 'doing nothing' that Odell prescribes, by exercising her attention in stubbornly unprofitable ways. But Mae's bosses at the Circle demand that even this activity must be shared and therefore made available as a commodity. ('Sharing is Caring' is the slogan they eventually elicit from her and adopt as their own.)

The wristbands of *The Circle* become known as 'ovals' by *The Every*, and they conform to the pattern of current wearable fitness tech in their attempts to colonise all unproductive time, but also in their emphasis on monitoring states of arousal. In *The Circle*, the campus doctor explains that the band 'allows you to know when you're amped or anxious'.[6] As Mae wears the band and monitors her temperature, her step count, her sleep, her heartrate, and so on, she feels 'a sense of great calm and control' with every new data point.[7] From 'amped and anxious' to calm and controlled, the Circlers are experiencing life in the preventative, algorithmically buffered zone of the quantified self, where no alarms are necessary because threats are always anticipated and erratic outputs can always be adjusted back to the norm with sufficient mindful attention. This is a neoliberal dystopia with very little coercion and no violence – just a wristband giving haptic feedback on a human wrist forever.

The Circle is careful to investigate the extent to which the characters' predicament is rooted in auto-exploitation, rather than in the direct monitoring of a threatening manager. For instance, when Mae is given the chance to choose an alert sound for a headset, she selects a computer-generated version of her own voice calling her name, a sound that she imagines as 'some older, wiser version of herself', a self that's 'purer', 'better', 'more indomitable'.[8] Rather than a boss bullying or threatening her, Mae is effectively disciplining herself. The self-administering quality of this activity recalls the philosopher Byung-Chul Han's description of contemporary power functioning through the ways people are persuaded to manage themselves:

> Under neoliberalism, the technology of power takes on a subtle form. It does not lay hold of individuals directly. Instead, it ensures that individuals act on themselves so that power relations are interiorized – and then interpreted as freedom. Self-optimization and submission, freedom and exploitation, fall into one.[9]

The underlying assumption of self-monitoring is that the self can be quantified, improved, and made more efficient or healthier through the augmented attention of a sensor and the algorithms that interpret its readings. This is a belief that Han has identified as a characteristic feature of contemporary neoliberal capitalism. For Han, we've entered a cultural moment in which we conceive of ourselves not as 'subjects' but as 'projects', finding a responsibility to attend to ourselves in ways which involve targeting areas for improvement. Identifying these cultural changes as a product of neoliberalism, which sees self-improvement as an obligation for increasing individual competitiveness, Han observes that we are 'always refashioning and reinventing ourselves' in a frenzy of 'compulsive achievement and optimization'.[10]

Body monitors are therefore not really about the body. In Han's terms, they're less biopolitical than psychopolitical. When the Lumo Lift buzzes you to stop slumping or the FitBit alerts you to your step goal, they are asking you to activate a monitoring attention that involves keeping track of yourself. You're being asked to imagine yourself as a project, ripe for improvement, and to activate your psychological processes of attentive, conscientious self-monitoring to do it.

The role of the wrist-worn body monitor as a specific marker of self-control is affirmed in Bernard E. Harcourt's observation that there is 'no longer any need for the state to force an ankle bracelet on us when we so lustfully clasp this pulsing, slick, hard object on our own wrist'.[11] *The Circle* and *The Every* capture the qualities of the fitness monitor that Harcourt identifies – those qualities that make it a desirable consumer object and a status symbol. But these devices also promise insight, with the implication that our feelings alone are unreliable or insufficient, or are too much for our attention to encompass. Rather than rely on the body's own cues to sleep or eat or drink or walk around or stretch, body monitoring devices imply that their users have (or ought to have) their minds on more absorbing things and will need reminding of their physiological needs.

At the end of *The Circle* there's an incredible scene which signals Mae's apparent conversion to the values of tech-giant's corporate values. Her best friend is in a coma, her body's internal processes visualised by monitoring of her vital signs, and Mae is frustrated by the impossibility of seeing right into her head – the very thing that we, as readers, have been doing to Mae (to a greater or lesser extent) for 500 pages. For Mae, the opacity of her friend's consciousness is 'an affront, a deprivation' in the face of The Circle's commitment to absolute transparency.[12] One reason I think that novels in particular have been fascinated by the wrist-worn tracker is because fiction too has an interest in monitoring and recording internal states in ways that take those feelings and sensations and re-present them to us. Fiction is, as the narrative theorist Dorrit Cohn has expressed it, committed to the dream of being able to read its characters' 'transparent minds'.[13] The novel entertains the fantasy that we might be able to see inside ourselves, transparently – this is what fiction and the FitBit have in common. The film adaptation of *The Circle*, starring Emma Watson as Mae, was unable to adapt the novel's complex shifting between narrative opacity and transparency. In a film there's little expectation that we might be invited to witness the explicit expression of a character's thoughts, or to see their interiority rendered transparently by a benevolently surveilling narrator: opacity is all.

Encircling the wrist, the body-worn alarm therefore promises absolute knowledge through its own monitoring attention. It's a version of the so-called 'completion' that The Circle also takes as its goal and which represents a future for the alarm in a form of total, perfect vigilance. Nothing will go unnoticed and nothing will be unexpected.

* * *

Should we therefore consider wearable self-monitoring devices a continuation or a repudiation of the logic of alarms? Even the alarm clock never claimed it would make you better at waking up, but many of these devices assert that they will teach you better habits, so that you regulate yourself without the need for their prompting. Instead of extending your attention (and thereby liberating you to do other things), these devices suggest they can train it.

One example is the Muse, which looks a bit like a call-centre headset, and describes itself as a 'multi-sensor meditation device' that can monitor breathing, heartrate, body movement and (distinctively) brain function. The device uses an EEG to give users 'real-time brainwave feedback' to teach 'the art of focus'.[14] Another similar device, the MIT Media Lab's AttentivU, has a note in its FAQs reassuring readers that it definitely does *not* read minds.[15] A dorky-looking, oversized pair of glasses, the AttentivU combines EEG readings with eye-tracking in order to monitor users' attention and alertness. The video made to promote the prototype device imagines situations such as listening to a boring university lecture or driving home after a night shift when you might appreciate the feedback from your AttentivU glasses to 'notify you of your attention habits'.[16] The device can give a sound or haptic vibration to remind you to stay alert. The publicity materials are all painstakingly framed around self-regulation rather than coercion – the scenario imagined is a student using the device to improve their own learning rather than a teacher using them for classroom discipline, for instance. But even the best-case scenarios still feel pretty dystopian: a world where, rather than having functioning public transport to get essential workers home from unsocial night working, they have to drive while fatigued with an occasional alert to stop them falling asleep at the wheel; a world where, rather than education being meaningful and engaging, students exhausted from off-campus work have to buzz themselves into participation. If I were speculating about the future, I'd rather imagine a better world than this.

Instead of using alarms to liberate attention, these devices have an ambition to track, measure, analyse, and alarm attention itself, with the aim of improving performance. Moreover, rather than the conventional alarm's urgent call to switch between opposing states of rest and arousal, self-monitoring devices operate by nudges, prompts and persuasions; they offer mild suggestions – a buzz, a blink – that barely even meet the threshold of a notification. They attempt to smooth over the jarring transition from one state of arousal to another, to avoid any extremes of feeling by charting a life that's well-balanced and optimised. They offer the calm without the alarm. And yet, instead of offering relief from the constant monitoring of the self or the environment – the prosthetic vigilance of the alarm – self-monitoring devices instead substitute a responsibility for constant mindful attention to the self. Where a conventional alarm tells you, 'Yes, your attention is not enough, but I'll take up the shortfall', the self-monitoring device says, 'Be mindful, be vigilant, be more attentive'. Attention monitoring devices like the Muse and the AttentivU take that logic to its extreme, by attempting to train their users into more acute focus and alertness. This is an anti-alarm philosophy, that puts its faith in a total attention that can prevent crises before they happen. If the alarm sounds, then the technology has failed.

As both a design philosophy and an approach to risk and safety, avoiding alarms is eminently sensible. 'Calm technology' is the name that's been given to the future products of the design philosophy that – among other tenets that promote a smoother human-machine interaction – eschews alarms. Amber Case offers a list of principles for calm technology, the first two of which are as follows:

I. Technology should require the smallest possible amount of attention.

II. Technology should inform and create calm.[17]

As we've established, the alarm is an inert technology which, when it functions correctly, uses the smallest possible amount of attention; its whole purpose is to minimise alarm and maximise calm. And in a world of increasing automation, as more processes that demand our attention could be taken care of by machines, there's a danger that alarms proliferate, as in the domestic chaos of beeps and buzzers in Frayn's *Alarms and Excursions*. The principles of calm technology exist to avert this kind of farce; to consider the context into which individual devices or programs emerge, rather than imagining them as the only demand on attention. Case observes that, too often, 'sharp and distracting' alarms are the product of arrogant engineers who believe that their technology is the most urgent, that their alerts should appear above all others.[18] Under the principles of calm technology, the urgency of the alert should match the urgency of the situation, so it's completely reasonable for a smoke detector to interrupt everything with a disruptive and shrill alarm, but not for a smart fridge that detects you're low on oat milk. One of Case's exercises for imagining calm technology is design prompt for a calm alarm clock, which might use gradual haptic or visual alerts make the transition from sleep to waking less jarring and unpleasant. As Case puts it, 'waking up doesn't have to mean being yelled at'.[19]

The Circle, like all good dystopias, takes a utopian ideal – this time of a world that prefers calm to alarm – and reveals its dystopian potential. When paired with the implacable logic of the attention economy, the laudable principles of calm technology are turned to sinister ends by the Circle's corporate demand for self-policing, for emotional regulation and auto-administration. Rather than overtly discipline their workers, the Circle uses modern management techniques that rely on workers feeling guilty, internalising messages about productivity, and controlling their own emotions to make their bosses feel good. The Circlers are expected to be as unalarmed – as controlled and calm – as the devices they use.

The future alarm, then, sheds as much of its alarm as it can, and amplifies the inertness and reliability that we've already recognised as significant properties of the alarm. We might say that the development of ambient, calm technology which offers a frictionless user-experience is very much informed by the logic of the alarm, because it affirms the idea of prosthetic vigilance, of technological objects taking on some of our attention. But this is also a future which finds the alarm's intrusiveness, its demands on attention, to be borderline violent: alarms are sharp, they're like being yelled at, they're disruptive – they're *alarming*. But to move into a utopian future, the technological possibilities of calm technology must be decoupled from the optimisation and self-administration of the attention economy.

* * *

The writer Joshua Cohen offers us a tidy aphorism for understanding the concept of the attention economy: 'When stimulated by capital, attention converts itself and responds as commodity'.[20] In an attention economy, attention feels like a resource that must be economised, but that doesn't mean that attention is, inevitably, a resource. The attention economy needs you to think that your attention is deficient, not just so that you'll buy a headband to buzz you if your brainwaves look distracted, but also because if you understand your attention as something that can be commodified, then it becomes your responsibility to optimise it or extend it in exactly the ways that benefit capital.

We've been given to understand attention as an individual commodity that can be counted and captured – something that can function as a unit in an economy – and which can therefore be economised. But what if we thought about attention in a different way? Yves Citton proposes the term 'attention ecology' as a recognition of the sense in which attention is transindividual, and emerges out of social relationships with each other. This aspect of attention can't be accounted for either in laboratory experiments that measure an isolated individual's reaction time or eye-tracking, or in economic conceptions of an attention economy that rely on individual rational actors stewarding out their attentional resources. Attention isn't the property of a single individual: as Citton explains it, attention 'calls for an exit from oneself'.[21] If the alarm extends some forms of attention, it also leaves others adrift. The forms of reflexive attention which participate in Citton's attention ecology are just one of those varieties of disregarded attention. The future alarm tends not to extend alertness outwards, exiting the self, but instead cultivates a monitoring vigilance that turns inwards, making attention ever more the property of an individual subject. *The Circle* serves as an early warning of some of the dangers of that hypertrophied attention.

At the beginning of this book we encountered Freud's concept of technology as a prosthetic addition to our own faculties. For Freud, the prosthetic technology causes us problems because we have not yet perfected ourselves through these devices, but for other thinkers who adopted the same prosthetic model for understanding technology, this dissatisfaction can be explained more by the way that technology extends and expands some faculties by blunting others. The media theorist Marshall McLuhan describes how overstimulation results from our interaction with technology and means some senses or faculties become numb or cut themselves off to protect themselves, while others are enhanced. He finds 'new ratios or new equilibriums' emerge when we use technology.[22] For alarms, this could mean some kinds of attention (vigilance, alertness) are extended with technology, in order that others (concentration or even the inattention of sleep) can develop. Moreover, this ratio or equilibrium might blunt or curtail other forms of attention that are neither extended nor enabled by the alarm, such as aesthetic attention, transindividual attention, or caring. We don't just project parts of ourselves outwards when we extend them, prosthetically, we also callous over or ignore others. Parts of our capacity for noticing, for attention, become less sensitive as others are extended or advanced. In the previous chapters we've seen how art can also make strange the familiar patterns of attention that vibrate through the alarm, and enable other forms of feeling and awareness to emerge.

The alarm has to be understood in relation to a persistent anxiety about human attention. It represents an attempt to master and control our wandering minds, even as it offers a way to manage the risks of an unpredictable world. Now, in the twenty-first century, at the same time as warnings about the future of the planet become ever more urgent, the alarms of the future seem intent on turning inwards to monitor attention itself. Much of this book has chronicled how alarms affirm the existing social relations of race, class, gender and property. But we have also seen the alarm serve as an object that rouses artists and writers into a new alertness to those forms of attention that the alarm ignores. In Lydia Davis's rewriting of the alarm archetype and KRS-One's re-voicing of the sound of the siren into lyric, in the heist movie's joyful capering through the art museum, and even in Francis Picabia's curious smashing of the alarm clock, we see attempts to convey forms of attention that open out beyond an individual extending their vigilance with an alarm. Art has taken the alarm's logic of attention and rerouted it into imaginary forms that become a drill or a rehearsal for what new forms of attention might be.

IMAGE CREDITS

Figure 7. Image from Augustus Pope's patent for an electro-magnetic burglar alarm, 1853. United States Patent and Trademark Office, US9802A.
Photo: United States Patent and Trademark Office.

NOTES

Introduction

1 Sigmund Freud, *Civilization and its Discontents* (London: Penguin Classics, 2002), trans. David McLintock, 30.

2 Ibid.

3 Mihaly Csikszentmihalyi, *Flow* (London: Rider, 1990).

4 'alarm, int., n., and adv.', *OED Online* (Oxford University Press, September 2021) www.oed.com/view/Entry/4547. Accessed 9 September 2021.

5 Ibid.

6 Christian Licoppe, 'The "Crisis of the Summons": A Transformation in the Pragmatics of "Notifications," from Phone Rings to Instant Messaging', *The Information Society* 26, no.4 (2010): 288–302.

7 My citations come from Anne Carson's plain and accessible translation of *Agamemnon* in Anne Carson, *An Oresteia* (New York: Faber and Faber, 2010), lines 205–34.

Chapter 1

1 The date here comes from Marius Hentea's *TaTa Dada* (Cambridge MA: MIT Press, 2012) because, in the accounts cited below from Hans Arp and Gabrielle Buffet-Picabia, each recalled a different date for the hotel meeting: Arp remembered 1917, while Buffet-Picabia thought it was February 1918.

2 Hans Arp, *Unsern Täglichen Traum* (Zurich: Arche, 1999), 64.

3 In her recollections, Picabia's wife, Gabrielle, was emphatic that the alarm clock was *not* stolen from the hotel room: 'The medium was an old alarm clock which we bought for a few cents and took apart. The detached pieces were bathed in ink and then imprinted at random on paper. All of us watched over the execution of this automatic masterpiece.' Gabrielle Buffet-Picabia 'Some Memories of Pre-Dada: Picabia and Duchamp', *The Dada Painters and Poets: An Anthology* Second Edition. Ed. Robert Motherwell (Cambridge MA and London: Belknap Press, 1989): 253–67, 266.

4 John Elderfield, *The Modern Drawing: 100 Works on Paper from The Museum of Modern Art* (New York: The Museum of Modern Art, 1983) 116.

5 Sarah Ganz Blythe and Edward D. Powers, *Looking at Dada* (New York: The Museum of Modern Art, 2006), 8.

6 William Rozaitis. 'The Joke at the Heart of Things: Francis Picabia's Machine Drawings and the Little Magazine 291.' *American Art*, vol. 8, no. 3/4, 1994, pp. 43–59. *JSTOR*, www.jstor.org/stable/3109171. Accessed 6 Apr. 2021.

7 Hans Arp, *Unsern Täglichen Traum* (Zurich: Arche, 1999), 64.

8 Viktor Shklovsky, 'Art as Technique', *Viktor Shklovsky: A Reader* (London: Bloomsbury, 2017) Ed. and trans. Alexandra Berlin. 80.

9 Ibid.

10 Westclox, 'Guardians of Sleep', *McCall's Magazine*, October 1929.

11 Franz Kafka, *The Metamorphosis*. Trans. Stanley Corngold (New York: Norton, 1996), 4.

12 *Franz Kafka: The Ghosts in the Machine* (Evanston, IL: Northwestern University Press, 2011), 194-5. Corngold and Wagner's reading of the health and safety diagram of the 'Accident Clock' in 'The Penal Colony' is also worth the attention of anyone interested in the ways workplace safety appears in Kafka's writing.

13 Benjamin Reiss, *Wild Nights: How Taming Sleep Created Our Restless World* (New York: Basic Books, 2017).

14 Jonathan Crary. *24/7: Late Capitalism and the Ends of Sleep* (London: Verso, 2013), 10–11.

Chapter 2

1 Business Week. 'The Blazing Market for Smoke Alarms'. 7 March 1977. MS55: Duane D. Pearsall Papers. Foundations of Fire Protection Engineering Collection, Digital WPI, Worcester Polytechnic Institute, Worcester, MA. Accessed 17 Sept 2021. https://digital.wpi.edu/show/ws859j082

2 You can view the archive on the Worcester Polytechnic Institute's website, under MS055: The Duane D. Pearsall Papers: https://digital.wpi.edu/collections/vx021h564. Many thanks to Arthur Carlson at WPI who provided the scans for the images in this book.

3   This description and the account I've given here comes from an excerpt of Pearsall's unpublished memoir, which is reproduced in a fascinating article by David A. Lucht in the *National Fire Protection Association Journal*. David A. Lucht, 'Where there's Smoke'. March/April 2013. MS55: Duane D. Pearsall Papers. Foundations of Fire Protection Engineering Collection, Digital WPI, Worcester Polytechnic Institute, Worcester, MA. Accessed 17 Sept 2021. https://digital.wpi.edu/show/3t945t533

4   Statitrol, 'Brochure for SmokeGard Smoke Detectors', 1 June 1972. MS55: Duane D. Pearsall Papers. Foundations of Fire Protection Engineering Collection, Digital WPI, Worcester Polytechnic Institute, Worcester, MA. Accessed 17 Sept 2021. https://digital.wpi.edu/show/f4752k239

5   National Fire Protection Association, '1975 ANSI/NFPA standards for installation, maintenance, and use of household fire warning equipment', 1975. MS55: Duane D. Pearsall Papers. Foundations of Fire Protection Engineering Collection, Digital WPI, Worcester Polytechnic Institute, Worcester, MA. Accessed 17 Sept 2021. https://digital.wpi.edu/show/2f75rb63h

6   *America Burning* has been digitised by the U.S. Fire Administration. National Commission on Fire Prevention and Control. *America Burning*. 1972. U.S. Fire Administration Publications. Accessed 17 September 2021. https://www.usfa.fema.gov/downloads/pdf/publications/fa-264.pdf

7   Don DeLillo, *White Noise* (London: Picador, 1985), 8.

8   Ibid., 262.

9   Ibid., 3.

10 Robert McCrum, 'Don DeLillo: 'I'm not trying to manipulate reality – this is what I see and hear' *The Guardian* 8 August 2010. Accessed 17 September 2021. https://www.theguardian.com/books/2010/aug/08/don-delillo-mccrum-interview

11 My ideas about home automation throughout this chapter (and especially the relationship between the 'push-button' home and the atom bomb) are indebted to Rachele Dini's wonderful book, *'All-Electric' Narratives: Time-Saving Appliances and Domesticity in American Literature, 1945–2020*. (London: Bloomsbury, 2021).

12 Don DeLillo, *White Noise* (London: Picador, 1985), 111.

13 Ibid., 156.

14 Ulrich Beck. *Risk Society: Towards a New Modernity*. Trans. Mark Ritter. (London: Sage, 2005).

15 Ibid., 36.

16 Statitrol 'Statifacts: SmokeGard Distributor Newsletter', September 1975. MS055.02.0020.00238: Duane D. Pearsall Papers. Foundations of Fire Protection Engineering Collection, Digital WPI, Worcester Polytechnic Institute, Worcester, MA. Accessed 17 Sept 2021. https://digital.wpi.edu/show/d504rn885

17 Michael Frayn, 'Alarms'. *Alarms and Excursions: More Plays than One*. (London: Methuen, 1998), 29.

18 Ibid., 15.

19 Ibid., 14.

20 Ibid., 11.

21 Ibid., 11-12.

22 Ibid., 13.

23 Greta Thunberg. "'Our house is on fire': Greta Thunberg, 16, urges leaders to act on climate'. *The Guardian* 25 Jan 2019. Accessed 17 September 2021 https://www.theguardian .com/environment/2019/jan/25/our-house-is-on-fire-greta -thunberg16-urges-leaders-to-act-on-climate

24 These are both slogans associated with Extinction Rebellion, including the handbook, *This is Not a Drill: The Extinction Rebellion Handbook* (Penguin, 2019).

Chapter 3

1 Augustus R. Pope. Improvement in Electro-Magnetic Alarms. US9802A, United States Patent and Trademark Office, 21 June 1853. *USPTO Patent Full-Text and Image Database.* Accessed 17 September 2021. https://pdfpiw.uspto .gov/.piw?docid=00009802

2 Edwin T. Holmes. *A Wonderful Fifty Years* (1917), 11. Digitised by HathiTrust. Accessed 17 September 2021. https:// hdl.handle.net/2027/hvd.hb0jjb

3 Ibid., 15.

4 Brandon LaBelle. *Acoustic Territories: Sound Culture and Everyday Life* (London: Continuum, 2010), 78.

5 David Foster Wallace. *Infinite Jest* (London: Abacus, 1997), 56.

6 Ibid., 56.

7 Ibid., 56.

8 Ibid., 448–9.

9 Ibid., 823.

10   Geoff Manaugh. *A Burglar's Guide to the City*, (New York: Farrar, Strauss and Giroux, 2016),18.

11   Ibid., 32.

12   Peter Brooks. *Reading for the Plot: Design and Intention in Narrative*. (Cambridge, MAL: Harvard University Press, 1992), 104.

13   Ibid., 104.

14   Julian Hanich. 'On Pros and Cons and Bills and Gates: The Heist Film as Pleasure.' *Film-Philosophy* vol.24, no.3 (2020): 304–20, 307.

15   Kim Newman 'The Caper Film', *The British Film Institute Companion to Crime*. Ed. Phil Hardy (Berkeley, CA: University of California Press, 1997): 70–1, 70.

16   National Security Advisor. 'Security in museums and galleries: advice for architects and planners'. *The Collections Trust* (2013). Accessed 17 September 2021. https://collectionstrust .org.uk/wp-content/uploads/2016/11/AdviceForArchitects AndPlanners_02.pdf

17   Daryl Lee. *The Heist Film: Stealing with Style*. (London: Wallflower Press, 2014), 10.

Chapter 4

1   Peter Sloterdijk. *Spheres: Volume 1 Bubbles, Microspherology*. Trans. Wieland Hoban (Los Angeles, CA: Semiotext(e), 2011), 500.

2   Ibid., 500.

3   Ibid., 500.

4 R. Murray Schafer. *The Soundscape: Our Sonic Environment and the Tuning of the World*, (Rochester, VT: Destiny Books, 1994), 332.

5 Ibid., 333.

6 'Loudest Siren'. *Guinness World Records.* Accessed 17 September 2021. https://www.guinnessworldrecords.com/world-records/loudest-siren

7 David Stall, a siren enthusiast, maintains a brilliant website with a huge number of scanned advertisements and instruction manuals for the Chrysler model sirens at victorysiren.org.

8 Jennifer Lucy Allan. *The Foghorn's Lament: The Disappearing Music of the Coast* (London: White Rabbit, 2021), 24.

9 Ibid., 24.

10 Ibid., 5.

11 Peter Sloterdijk. *Spheres: Volume 1 Bubbles, Microspherology*. Trans. Wieland Hoban (Los Angeles, CA: Semiotext(e), 2011), 500.

12 Michael Bull. *Sirens* (New York, NY and London: Bloomsbury, 2020), 37.

13 'Emergency Alerts'. *GOV.UK.* Accessed 26 July 2022. https://www.gov.uk/alerts.

14 KRS-One. 'Sound of da Police'. *Spotify*. Accessed 21 September 2021. https://open.spotify.com/track/3Y6XWs8xMlCngyIxNOFnsp?si=23abec00bd3149d2.

15 Richard Wright. *Native Son* (London: Vintage, 2000), 288.

16 Jennifer Lee Stoever. *The Sonic Color Line: Race and the Cultural Politics of Listening.* (New York: New York University Press, 2016), 224.

17 Ibid., 225–6.

18 Richard Wright. 'How Bigger was Born'. *Native Son* (London: Vintage, 2000), 25.

19 Richard Wright. *Native Son* (London: Vintage, 2000), 290.

20 Simone Browne. *Dark Matters: On the Surveillance of Blackness*. (Durham, NC: Duke University Press, 2015), 21.

21 'Caution! Colored People of Boston' handbill from 1851, reproduced in Simone Browne, *Dark Matters: On the Surveillance of Blackness*. (Durham, NC: Duke University Press, 2015), 23.

22 Claudia Rankine. *Citizen: An American Lyric*. (London: Penguin, 2014), 134.

23 Ibid., 15.

24 Ibid., 10.

25 Ibid., 151.

26 Ibid., 151.

27 Claudia Rankine and John Lucas. 'Situation 6 (Stop and Frisk)'. *PBS NewsHour on Youtube*. 4 Dec 2014. Accessed 21 September 2021 https://www.youtube.com/watch?v =kN5aYIrc2J8

28 Claudia Rankine. *Citizen: An American Lyric*. (London: Penguin, 2014), 107–8.

29 For a much fuller reading of repetition and racial trauma in the poem, see Mary Jean Chan. 'Towards a Poetics of Racial Trauma: Lyric Hybridity in Claudia Rankine's *Citizen*'. *Journal of American Studies*, vol. 52, no.1: 137–63,

30 Rankine, 105.

31 Shermaine M. Jones. '"I Can't Breathe!": Affective Asphyxia in Claudia Rankine's *Citizen: An American Lyric*'. *South: A Scholarly Journal*, vol. 50, no. 1, 2017: 37–46.

32 Sandeep Parmar. 'Not a British Subject: Race and Poetry in the UK', *LARB* 6 December 2015. Accessed 21 September 2021 https://lareviewofbooks.org/article/not-a-british-subject-race-and-poetry-in-the-uk/.

Chapter 5

1 Sara Ahmed. *The Cultural Politics of Emotion.* (London: Routledge, 2007), 11.

2 Ibid., 92.

3 Stephen R. Dixon, Christopher D. Wickens, and Jason S. McCarley. 'On the independence of compliance and reliance: Are automation false alarms worse than misses?' *Human Factors* vol. 49, no.4 (2007): 564-572.

4 Bill R. Hollifield and Eddie Habibi. *The Alarm Management Handbook: A Comprehensive Guide.* (Houston TX: PAS, 2016), 118.

5 Lydia Davis. 'Fear'. *The Collected Stories of Lydia Davis.* (London: Penguin, 2011), 258. You can read the whole story online at *Conjunctions* magazine: http://www.conjunctions.com/print/article/lydia-davis-c24

Chapter 6

1 *Ōura.* Accessed 21 September 2021. https://ouraring.com/

2 Amazon Technologies, Inc. 'Ultrasonic Bracelet and Receiver for Detecting Position in 2D Plane'. US9881276B2, United

States Patent and Trademark Office, 30 January 2018. https://pdfpiw.uspto.gov/.piw?Docid=09881276; Amazon Technologies, Inc. 'Wrist Band Haptic Feedback System'. US9881277B2, United States Patent and Trademark Office, 30 January 2018. https://pdfpiw.uspto.gov/.piw?Docid=09881277

3 Dave Eggers, *The Circle* (London: Penguin, 2014).

4 Ibid., 195.

5 Jenny Odell, *How to Do Nothing: Resisting the Attention Economy* (New York: Melville House, 2019).

6 Eggers, 154.

7 Ibid., 194.

8 Ibid., 232, 331.

9 Byung-Chul Han, *Psychopolitics: Neoliberalism and New Technologies of Power*. Trans. Eric Butler. (London: Verso, 2017), 28.

10 Ibid., 1.

11 Bernard E. Harcourt, *Exposed: Desire and Disobedience in the Digital Age* (Cambridge, MA: Harvard University Press, 2015), 124.

12 Eggers, 491.

13 Dorrit Cohn, *Transparent Minds: Narrative Modes for Presenting Consciousness in Fiction* (Princeton University Press, 1984).

14 *Muse.* Accessed 22 September 2021. https://choosemuse.com

15 'Terminology'. *AttentivU.* Accessed 22 September 2021. https://www.attentivu.com/terminology

16 *AttentivU.* Accessed 22 September 2021. https://www.attentivu.com/

17 Amber Case. *Calm Technology: Principles and Patterns for Non-Intrusive Design* (Beijing: O'Reilly, 2015), 48.

18 Ibid., 83.

19 Amber Case, 'Exercises in Calm Technology'. *Calmtech.com*. Accessed 22 September 2021, https://calmtech.com/exercises.html.

20 Joshua Cohen. *Attention!: A (Short) History* (London: Notting Hill Editions, 2013), 182.

21 Yves Citton. *The Ecology of Attention*. Translated by Barnaby Norman. (Cambridge: Polity, 2017), 78.

22 Marshall McLuhan, *Understanding Media: The Extensions of Man* (Cambridge MA and London, MIT Press, 2001), 45.

INDEX

OBJECT LESSONS

Cross them all off your list.

9781501358159

9781501353277

9781501348716

9781501353024

9781501344350

9781501361906

"Perfect for slipping in a pocket and pulling out when life is on hold."

– Toronto Star

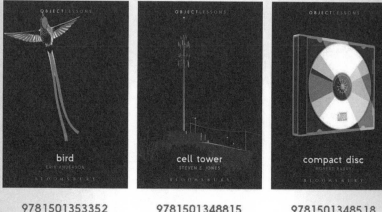

bird 9781501353352	cell tower 9781501348815	compact disc 9781501348518

ocean 9781501348631	high heel 9781501325991	hood 9781501307409